Experiments
with
Integrated Circuits

Experiments with Integrated Circuits
The IC Electronics Toolbox

RICHARD J. HIGGINS
University of Oregon

PRENTICE-HALL, INC., *Englewood Cliffs, N.J.* 07632

Material in the following figure used courtesy Advanced Micro Devices: Appendix I.

The following figures used courtesy Analog Devices: 14.3, 20.2, 21.1, and material in Appendix I.

The following figure is used courtesy Global Specialities Corp.: P6(b).

The following figure is used with permission of Richard J. Higgins: 26.1.

Material in the following figure used courtesy Intel: Appendix I.

The following figure is used courtesy Kiethley Instruments: P6(a).

Material in the following figure used courtesy Motorola: Appendix I.

The following figures are taken from Higgins, *Electronics with Digital and Analog Integrated Circuits.* Copyright © 1983 by Prentice-Hall, Inc.. Cover figure, 1.1, 1.3, 6.2, 6.3, 7.1, 7.2, 7.4, 7.5, 8.1, 8.2, 8.3(c), 8.4, 9.1, 14.1, 14.2, 15.1, 16.2, 17.1, 17.2, 17.3, 17.4, 17.5, 19.1, 19.2, 20.1(b), 22.1, 22.2, 22.3, 22.4, 23.1, 23.2, 23.3, 23.4, 23.5, 24.1, 24.2, 25.1, and 27.1

Material in the following figure used courtesy RCA: Appendix I.

Permission to reprint Figures 27.2, 27.4, and material in Appendix I granted by Signetics Corporation, a subsidiary of U. S. Philips Corp., 811 E. Arques Avenue, Sunnyvale, CA 94086.

The following figure is used courtesy Tektronix, Inc.: P6(c).

The following figures are used courtesy Texas Instruments Incorporated: 6.1, 7.3, 8.3(a), 10.1, 11.3, 13.1, 13.2, and material in Appendix I.

© 1983 by Prentice-Hall, Inc., Englewood Cliffs, New Jersey 07632

All rights reserved. No part of this book may be reproduced, in any form or by any means, without permission in writing from the publisher.

Printed in the United States of America

10 9 8 7 6 5 4 3 2 1

ISBN 0-13-295527-X

Prentice-Hall International, Inc., *London*
Prentice-Hall of Australia Pty. Limited, *Sydney*
Editora Prentice-Hall do Brasil, Ltda., *Rio de Janeiro*
Prentice-Hall Canada Inc., *Toronto*
Prentice-Hall of India Private Limited, *New Delhi*
Prentice-Hall of Japan, Inc., *Tokyo*
Prentice-Hall of Southeast Asia Pte. Ltd., *Singapore*
Whitehall Books Limited, *Wellington, New Zealand*

CONTENTS

Notes to the Student
      Selecting Experiments
      Basic Assumptions and Strategies
      Lab Procedures
      List of Integrated Circuits - The Basic Kit
      Suggestions for Special Project Labs

Notes to the Instructor
      Wiring up Circuits
      Testing Circuit Functions
      Selecting Components for the Electronics Toolbox
      Selecting Instruments for the Electronics Lab

<u>Experiment</u>   <u>Title</u>

1. Discrete Component Electronic Circuits and Measurements 1-1

2. Count by Two's: Binary Numbers 2-1

3. Gating Circuits 3-1

4. Boolean Algebra in Logic Design 4-1

5. Digital Decoding, Multiplexing, and Sequencing 5-1

6. Flip Flops 6-1

7. Counters 7-1

8. Digital Wave Generation and Waveshaping 8-1

9. A Counter Measurement of Elapsed Time 9-1

10. Shift Registers 10-1

11. Memory 11-1

12. Binary Adders 12-1

13. Putting it together: a TTL MSI 4-bit Microcomputer 13-1

14. Digital to Analog Conversion 14-1

15. Analog to Digital Conversion by the Dual-slope Method 15-1

16. Analog to Digital Conversion by the Successive Approximation Method 16-1

17. Basic Operational Amplifier Circuits 17-1

18. Operational Amplifier Applications in Simulation 18-1

19. Analog Computer Solutions of the Damped Harmonic Oscillator 19-1

20. Log Amps 20-1

21. Multipliers 21-1

22. Active Filters 22-1

23. Oscillators 23-1

24. Function Generators 24-1

25. Low Level Signals I: High Performance Op Amps 25-1

26. Low Level Signals II: Lock-In Amplifier 26-1

27. Phase-Locked Loops 27-1

APPENDICES
Pin Diagrams for Common IC's
How to Identify Components: R, C, et al.
Sources of Supply

NOTES TO THE STUDENT

SELECTING EXPERIMENTS

1. Discrete Component Electronics
   Do this one if you are unfamiliar with wiring up practical circuits (both passive and active), or are not at home with basic electronic test instruments.

DIGITAL EXPERIMENTS
2. Count by Two's
   Simple concepts, but introduces you to the basic hardware for breadboard circuit contruction.

3. Gating Circuits
6. Flip flops
   Both are absolutely basic. Everyone should do them.

4. Boolean algebra
5. Decoding
   Fundamental enough, but of interest mostly to the computer scientists or future hardware designer.

7. Counters
   Concepts easily picked up in use. Could be skipped by most people, but here for completeness.

8. Waveshaping
9. Elapsed Time
   Fundamental for scientists. Number 8 deals with data communications. Number 9 deals with digital measurement.

From here on in the digital labs, there are two parallel branches which apply the fundamentals learned above. Most people will prefer to select one of the two following sequences:

Scientific Instrumentation
14. Digital to Analog
15. Analog to Digital (Dual Slope)
16. Analog to Digital (Successive Approximation)

Computer Hardware
10. Shift Registers
11. Memory
12. Binary Adders
14. TTL Microcomputer

ANALOG EXPERIMENTS

17. Basic Operational Amplifier Circuits.
    Essential; key op amp principles and practical realizations of mathematical operations.

Control and Simulation group: pick 18 or 19.
18. Operational Amplifier Applications in Control: the Voltage Regulator.
19. Analog Computer Solutions of the Damped Harmonic Oscillator. Simulation of systems via their differential equation.

Nonlinear Circuits:z Pick 20 or 21.
20. Log Amps (Wide dynamic range amp)
21. Multipliers (Nonlinear mixing of signals)

Waveshaping Group: Pick 22, 23, or 24
22. Active Filters (high-pass, low-pass, band-pass)
23. Oscillators and Instability (phase-shift, twin-tee, and tuned-circuit oscillators)
24. Function Generators (Square and triangle waves with voltage-controlled frequency)

The Measurerments Group: Pick 25, 26, or 27
25. Low Level Signals and High Perfomance Op Amps (instrumentation amplifier circuit)
26. Lock-in Amplifier (Coherent or phase-sensitive detection)
27. Phase-locked Loops (Tracking oscillator for modulation, data transmission, and harmonic generation)

BASIC ASSUMPTIONS AND STRATEGIES

Background Assumed. No formal electronics prerequisites. Students without previous hands-on electronics lab experience should do experiment 1. Others may skip it and procede to the IC experiments.

Black Box Approach. Little emphasis will be placed on detailed electronic circuit analysis or design, because integrated circuits make this less necesssary. On the other hand, enough time will be spent on exploring subtleties of circuit operation to know how to select IC's intelligently. We encourage you to make extensive use of equivalent circuit models and fundamental theorems which allow you to characterize the operation of a circuit, treating it as a black box, i.e., without looking inside the box.

Selecting Experiments. In both the digital and the analog sections, there are a few basic experiments which nearly everyone should go through. After successful completion of those, considerable freedom is available to select experiments most closely matched to needs and interests. Students in the sciences will often prefer to select measurement-oriented experiments. Students in computer science will usually put their priorities on the hardware aspects of computers. Consult the page Selecting Experiments for specific guidelines.

Jump Right In. We will often use an IC and only later come back to understand how it really works. For example, experiment 2 on binary numbers uses IC counters, though the background in flip-flops and counters doesn't come until later labs. Experiments 14-16 make use of op amps even before the basic op amp labs. This lets you get right to interesting applications without an impossibly long list of basic circuit labs. You may feel uncomfortable about it; this is normal. Trust the black box approach.

Expectations. The experiments have been designed to be done one-per-week in a three-hour lab period, with 8 (quarter system) or 10-12 (semester system) being typical expectations. Some labs take longer (6: flip flops; 18: basic op amp circuits, for example) and are so fundamental that they deserve extra time. The instructor should schedule extra hours or even unsupervised open hours (once you have proven you will not damage the equipment or yourself). A list of <u>project labs</u> is also given to suggest end-of-term final projects which lend themselves naturally to term paper or open house presentation.

LAB PROCEDURES

<u>The Kit</u>. We have found it best to issue to individuals a kit containing basic integrated circuits, passive components, a breadboard, and simple test instruments. The individual(s) assigned a given kit thereby has guaranteed access but also responsibility for the kit. It is inevitable that you will burn out some IC's, but be aware that even though gates at a few cents apiece are essentially free, other components may be more expensive (e.g., D/A converters at $10). Test instruments have become small enough to fit inside the kit or, unfortunately, even in a pocket. It is essential to see that test instruments do not wander away. The kits are to be returned and checked before course grades are assigned.

<u>Solo Performance</u>. There should be no lab partners. It is a goal of the lab course for everyone to have the chance to fumble through wiring up circuits and getting test instruments to work. This is called "hands-on experience." Kits and equipment can be shared but in different time periods. On the other hand, an ambitious final project often benefits from a collaboration. Consult your instructor on sections marked <u>optional</u>. We have used them to provide more depth for grad students, for example, or to distinguish A-level performance.

<u>Your Lab Notebook</u>. You need not repeat explanations printed in this lab book. However, your lab notebook should include a clear statement of what you set out to do, what you observed, any unusual happenings and their explanation. If you let your curiosity go free it will add depth to the experiments and build some good habits of scientific investigation. The lab notebook should therefore mimic what any careful scientist records, though with adequate clarity for someone else to read. Do not, however, take data on scrap paper during the experiment and recopy it later into your lab notebook.

Hardware Notes

1. Pins on IC's and plug-in test modules (<u>Outboards</u>, for example) break easily. Do not remove them by wiggling back and forth. Pull them straight out, using a small screwdriver underneath for leverage.

2. IC's burn out easily, especially for the novice. Consult your instructor to get replacements. On the other hand, there is no need for expensive IC's or test instruments from your kit to end up missing.

3. What to do with unused gate inputs? Think about it and try it. Do unused TTL inputs float high or low?

4. How to destroy an IC:
   a. Power leads reversed on breadboard;
   b. IC plugged in backwards (dual-in-line or DIP package) or rotated (circular package);
   c. Signals applied without power being supplied to IC (likely to kill TTL; will surely wipe out CMOS).

5. What to do if you suspect an IC has been destroyed by you or by a previous user?
   a. If the package has several functionally equivalent circuits, try another.
   b. Leaving the wiring unchanged, substitute an identical IC. But if that kills a previously good IC, check the wiring again.

6. If you discover that one circuit on a multi-circuit IC has burned out, do everyone a favor and bend or clip off the leads so it can never be used again.

LIST OF INTEGRATED CIRCUITS NEEDED

555	Timer
556	Voltage-controlled oscillator (optional)
565	Phase-locked loop
741	Operational Amplifier (4 or more)
7400	Quad 2-input positive NAND gates
7402	Quad 2-input positive NOR gates
7404	Hex inverters
7405	Hex inverters, open collector
7408	Quad 2-input positive AND gates
7410	Triple 3-input positive NAND gates
7432	Quad 2-input positive OR gates
7442	4 to 10 line decoder
7451	Dual 2-wide, 2-input AND-OR-Invert gate
7474	Dual D-type edge-triggered flip flop
7475	4-bit bistable latch
7476	Dual JK flip flop with clear and preset
7486	Quad 2-input Exclusive OR gates
7489	65 bit random-access memory
7490	Decade counter
7493	4-bit binary counter
74121	Monostable multivibrator (one-shot)
74126	Quad bus buffer gates with Tri-state outputs
74150	16 to 1 line multiplexer (4 address-bits)
74154	4 to 16 line decoder or 16 to 1 demultiplexer
74193	Synchronous 4-bit up/down counter
74194	Shift register, parallel in, parallel out
8038	Function generator (sine, square, triangle)
AD504	Op amp, low voltage drift
AD506	FET-input op amp (substitution: any BiMOS or BiFET)
AD533	Analog multiplier
AD7533	Digital to analog converter (or equivalent, such as AD7520 or DAC-100)
AM2502	Successive approximation register
CA3140	BiMOS op amp (substitution: TL080)
DG200	Analog switch (substitution: AD7510; DG200)
TL080	BiFET op amp (or equivalent, e.g., LF356 or AD547)

SUGGESTIONS FOR IC ELECTRONICS PROJECT LABS

Dual slope DVM and digital thermometer. Combine an integrated dual slope A/D chip (e.g., Motorola MC14433 or Intersil 7106/7107) with a 4 digit LCD display to make a digital voltmeter. Or, use complete A/D plus LCD available inexpensively in an evaluation kit (e.g., Intersil 7106EV). Then add an inexpensive linear semiconductor transducer (e.g., Analog Devices AD 590 or National LM 335Z). The AD 590's output is precisely 1 uA/K! Other transducers are readily digitized in this manner. For example, pressure (e.g., National's semiconductor pressure transducer series), position (any slide or rotary potentiometer), etc.

Frequency/time as universal measurement variables. Combine an inexpensive multi-digit counter/timer IC (e.g., Intersil 7045 or 7205 or 7215 timer or 7226 frequency counter) with an inexpensive 4-digit LCD display to make a frequency counter or time interval measurement device. Or, use complete timer/counter plus LCD available inexpensively via an evaluation kit (e.g., Intersil 7045EV or 7205EV or 7215EV stopwatch or 7226EV counter). The freedom to select the range of time interval or clock frequency lends flexibility to the measurement. For example, a flute player might want to explore not only tuning intervals but also how much frequency shift occurs during vibrato. This involves some tricky sampling, and consideration of tradeoffs between measurement speed and resolution. Someone with an anemometer might want to design a wind velocity measuring instrument calibrated directly in kilometers per hour. A normal anemometer's rotating vanes drive a slotted wheel which passes between an LED and a photodetector. The output is a more-or-less binary square wave (some waveshaping will be needed) whose frequency is a measure of wind speed.

A/D input for your personal computer. Interface a dual-slope A/D IC for high resolution measurements. The Motorola MC 14433 or Intersil 7106 lend themselves readily to this, since they have the binary outputs and control lines needed. These high resolution devices normally transmit data digit by digit, strobed out of an output register. You will wire one port of the computer to receive the digits. The program needs to control <u>when</u> digits are received, and format them back from this "$\overline{BCD}$ bit-serial, byte parallel" format. If the speed of this method is too slow, use a successive approximation A/D. The successive approximation logic is available on one chip (e.g., Advanced Microdevices AM2502). Alternatively, the combination of successive approximation logic, D/A, and interfacing logic are now available complete on one IC (e.g., Analog Devices AD570, or, with handshaking logic, the AD673).

High speed arithmetic. Evaluate one of the high speed multiplier or multifunction arithmetic chips. Choose your chip carefully; the less expensive ones may be slower (e.g., National MM57109) or, if not, require <u>you</u> to provide a lot of external logic (e.g., TRW MPY-8). On the other hand, do-everything chips are expensive (e.g., Advanced Microdevices AM9511). Design an add-on to your personal computer which strobes in the data to be calculated with, and provides the control logic to let the computer know when it is ready to send the answer back.

Smart A/D board with local memory. This is equivalent to a digital scope or <u>waveform digitizer</u> (e.g., Biomation). The circuit encodes a <u>set</u> of samples (e.g., audio tone) for later processing by a slower microcomputer.

Candy machine logic (a gate analysis exercise). Consider the choices a candy machine has to make: money received? which item selected? change needed? Design a gating network to ask and answer the questions.

Digital electronic music for guitar or other non-keyboard instrument. Use a phase-locked loop (try CD 4046, CMOS) to measure the pitch of the note being played. Divide that frequency down by 32 or more by a counter. Let the counter outputs strobe through the address range of a RAM (try MCM 6810L, 128x8) which contains the waveform to be played. The RAM output is converted to a voltage waveform by a D/A.

Poly-rhythmic metronome. Design a counter circuit which <u>simultaneously</u> generates on its 8 output lines pulses of 1 to 8 beats per measure. The frequency of the clock is set 840 times higher than the metronome's beat, since 840 is the least common multiple. Use a 555 timer and make the frequency variable over about a 5 to 1 range to imitate a metronome's range. Use CMOS (low power) counters wired as divide-by-N circuits, and feed the outputs to LED's.

Hamming error correction. Construct in hardware the Hamming algorithm, which works on data in 4-bit wide nibbles and detects 1 or 2-bit errors, <u>corrects</u> 1-bit errors, and outputs error code information. Each 4-bit word to be transmitted has appended to it another 4-bits, the Hanning code for that word, generated in a lookup table in RAM. The 8-bit received word is checked for errors with a parity generator IC and some logic. Reference: Byte magazine, Feb. 1979, pp. 180-182.

Linear light measurements. Use a Si solar cell whose output is linearly proportional to light power when the cell is operated in a short circuit current-source mode. Use an op amp current-to-voltage circuit to make a measurable signal.

Light controller. Feedback circuit which adjusts a light bulb to keep measured light intensity constant independent of fluctuations in ambient light. Ref: Text Fig. 13.10.

Control Systems. Control of physical systems using an op amp and feedback. Stability. The language of control theory: proportional, rate, reset. Applications could be mechanical, thermal, optical...

Nonlinear Simulations: Bouncing Ball. Diodes in the feedback loop create the restoring force of a hard floor.

Nonlinear Op Amp Circuits. Diodes in the feedback loop to create: precision ac voltmeter; threshold detector; precision clipper; arbitrary function generator; precision switch; absolute value circuit; backlash simulator.

Nonlinear Simulation: Foxes and Rabbits. Classic host-parasite problem of ecology; lack of ecological complexity brings instability. Multiplier couples two linear systems.

Quantum Mechanics on the Analog Computer. Visually satisfying analog simulation. One sees the importance of eigenvalues: they give the only solutions which do not blow up. Simplest example: square well.

The Ubiquitous 555 Timer. An analog-digital hybrid, with time delays from microseconds to hours. Use the retriggering mode to make clever applications such as a burglar alarm when a light beam is interrupted.

Brain Waves: Alpha and all that. Use a sensitive op amp to observe brain waves on a scope. Active filter Selects particular frequency components characteristic of several modes of brain activity. Close to feedback loop to make "biofeedback."

Linear Electronic Thermometer. Use commercial transducer such as AD590, whose output is precisely proportional to absolute temperature.

Phase Locked Loop Applications With Tape Recorders. Using an inexpensive cassette tape recorder, construct the equivalent of an expensive fm tape recorder which can record analog signals all the way down in frequency to dc! A modification of the same circuit gives a means of off-line program or data storage for digital communications.

Slew-rate limiting filter. Nonlinear response to noise spikes while keeping linear response to audio signals of interest. Ref: Text Section 19.2.1.

MISCELLANEOUS PROJECT SUGGESTIONS

Digital data acquisition: <u>multiplexed</u> A/D
Pulse counting for photons or particles
Digital filters for signal processing
Tracking A/D converter
Companding A/D converter
Phase modulation (delta modulation) A/D for digital communications and digital audio
Signal averaging
Correlation of two signals
Analog shift registers for audio or video signal processing
Digital electronic music
Microcomputer interfacing (your pet project)

NOTES TO THE INSTRUCTOR

As the preface to the text indicates, learning electronics not only gives one a feeling of power but is also intended to be fun. The fun should include the instructor, no matter what his or her previous electronics background. The adage "You can't teach an old dog new tricks" does not apply to someone learning how to use IC's. We include in this section some hints for setting up the minimal electronics lab, so you can get these experiments going inexpensively and without an unreasonable burden on your time.

But aren't some of these TTL IC's obslete? The 4-bit TTL IC's are obsolete in the sense that newer more-bits-per-package MOS and CMOS devices are available. While 8 bits is more standard in microcomputers, we use 4-bit-wide TTL devices in the beginning digital circuits because they are:
- (a) enough to represent a hexadecimal or decimal digit;
- (b) interface easily to inexpensive tutorial test modules;
- (c) quick to wire and cycle through the possible states;
- (d) cheap and readily available anywhere;
- (e) TTL MSI which is rugged enough to withstand much more abuse than MOS devices.

Once the principle of random-access memory is demonstrated by wiring up a 64-bit 7489, the extension to more typical IC's is readily understood, without the work of wiring it up or trying it out in the crude breadboard stage. On the other hand, 4-bit wide data has poor resolution. A sine wave stored in 4-bit format is not very accurate. Most microcomputer interfacing is done to 8-bit ports, and ASCII communications involves 8-bit data. Therefore, where appropriate, 8-bit alternatives are included, for example in experiments 10 (shift registers), 11 (memory) and 14 (D/A). Higher resolution routes are standard in experiments 9 (elapsed time), 15 (dual slope A/D), and 16 (successive approximation A/D).

Wiring Up Circuits

To what extent should circuits be made available to students prewired, or should students wire up everything? We find a compromise works best. Students gain an all-important mastery from having to search out a lack on electrical continuity, a wiring error, a source of power line noise, an IC plugged in backwards, a missing ground, a ground loop. Use reliable but quick wiring breadboards; examples are shown in Figure P1. On the other hand, too much time spent wiring and too little spent measuring is the wrong balance. A list is given at the end of this section of circuits which we have found it best to prewire for students. For circuits which you prebuild and for student projects, we recommend that you make available a wire-wrap tool and appropriate wire-wrap hardware, and/or a tool for point-to-point soldering.

Circuits Which We Recommend You Pre-wire for Students

Some are not that difficult, but add an unnecessary complication which lends little knowledge. Some, however, are just downright difficult to wire up.

Experiment	Circuit	Figure
1	Transistor amplifiers	1.3
6	Glitch detector	6.4
15	Dual-slope A/D converter	15.1
16	Successive approx. A/D (TTL-MSI chip version)	16.1
27	Lock-in amplifier	27.1 or 27.2
28	Phase-locked loop	28.1

Testing Circuit Functions

For digital circuits, we recommend that standard circuit test functions be made available either on a "logic trainer" box or on modules which plug on to the breadboard. The basic list is:

    logic switches, 4, debouncing not essential
    debounced pulsers, 2
    clock, variable frequency (1 Hz to 1 MHz)
    binary display lights, normally 4 or 8 LED's
    BCD numeric display, LED or LCD. One digit will do,
        though four digits are required for measurements.

The philosophy is similar for analog circuits, with a ± 15 V power supply and appropriate breadboard bus strip connections. Examples of logic trainers and test modules are given in the text, Fig. 2.6. See also the Appendix Sources of Supply. Test modules shown on layout diagrams in Labs 2-4 are the "Outboard" series from E & L Instruments, Derby, Connecticut.

Selecting Components for the Electronics Toolbox

These practical guidelines are a subset including only the kinds of components most commonly used in IC circuits.

Resistors. Two things matter: the material and the power dissipation. For routine circuit use, the carbon composition resistor is common. But carbon is a semiconductor, so the resistance varies with temperature. To avoid this, use a metal film resistor, whose resistance varies much more slowly with temperature.

Type	Temperature coefficient of resistance, percent/(deg C)
Carbon composition	0.1
Metal film	0.01

How can you tell them apart? Use catalog specifications to be sure, though you will soon recognize them by their shape. If the precise value of the resistance matters, use a precision resistor. Normal resistance tolerances are 5-10%, precision resistors are easily obtainable with tolerances of

1%, and 0.01% tolerance can be obtained with extra delay and expense.

If smoke comes out of the resistor, the <u>power</u> <u>dissipation</u> capability is not high enough. Typical circuit resistors can dissipate only 1/4 W. High power resistors are much larger and have cooling fins [Fig. P2(b)].

You will also use a variable resistor or <u>pot</u>, short for potentiometer. It's a misnomer; the word comes from a method of measuring a voltage using a variable resistor plus reference voltage ("potential meter"). The most common pot in IC circuits, called a <u>trim</u> <u>pot</u>, adjusts or "trims" the value of some variable. Trim pots are made to plug neatly into the same pin spacing as IC circuits [Fig. P2(c)].

Capacitors. Most capacitors are two strips of metal foil wound up with a thin sheet of insulating dielectric in between. There are upper limits to the capacitance value which can be obtained this way in a reasonable size package. <u>Electrolytic</u> capacitors have a conducting electrolyte (previously liquid; solid electrolytes are now available) between the metal plates [Fig. P3(a)]. The dielectric is an oxide layer on one plate, formed by electrolytic action. Because the oxide is thin, the capacitance is large. Most electrolytic capacitors are <u>polarized</u>: one side must be connected to a positive voltage relative to the other side. Electrolytic capacitors also have large <u>dissipation</u> <u>factors</u>: they allow some current to leak through, acting as a resistor in parallel with the capacitor. Electrolytic capacitors are therefore only used where their large C-values are an asset and their disadvantages pose no problem. Electrolytics for power supplies are large and bulky. Tiny 1 uF electrolytics are handy on each circuit board for <u>decoupling</u> each board from current fluctuations in others.

The most common non-electrolytic capacitors in IC circuits are the <u>ceramic</u> <u>disk</u> (so called because of the dielectric material and shape) and <u>plastic</u> <u>film</u> (usually "mylar" or "polyester") [Fig. P3(b)]. Capacitor parameters include:

(a) Tolerance: deviation from specified value; $\pm 10\%$ tolerance is typical unless you specify otherwise.

(b) Dissipation: ratio of capacitive reactance to

resistance.

(c) Frequency variation: The capacitance value will decrease at very high frequencies.

(d) Distortion: If the capacitance value changes with applied voltage (due to deformation of the dielectric), the nonlinearity will introduce harmonic distortion into a sine wave.

(e) Maximum voltage: Dielectric breakdown is the most common cause of capacitor failure. An electrolytic capacitor subjected to long-term overvoltage can fail explosively.

Diodes and Transistors. The first criterion is power dissipation. Will the device be handling signals at low power level, or driving large currents? The mechanical and thermal design varies accordingly [Fig. P4]. Devices which must handle more than a few watts are designed for attachment to a metal plate or heat sink for cooling. The second criterion is speed. How fast can the output respond to changes in the input? Most transistors can handle transitions in the sub-microsecond range easily. Transistors and diodes designed for power supply use have larger device area and are often more limited in speed as a result of junction capacitance. Finally, for analog applications, linearity is important, and is often specified by quoting a harmonic distortion limit. Though transistors are optimized either for switching or for linear amplification, the division is not absolute, and a switching transistor can make a respectable linear amplifier. (See Appendix for a selection of specifications.)

Batteries and Voltage References. For most electronics, a power supply is preferable to batteries. Digital logic experiments can be done with a 6 V "dry cell" lantern battery, but the results become erratic as the voltage falls. But battery power finds a place in portable instruments, or to make a simple floating ripple-free reference voltage. Battery type is usually specified by the plate and/or electrolyte materials. The dry-cell flashlight battery is Carbon-Zinc or C-Zn; a car battery is Pb-PbO. The term alkaline battery specifies a C-Zn battery with higher conductivity electrolyte and therefore higher power output. Battery types differ also in voltage stability during the discharge cycle, an important parameter in electronic applications [Fig. P5].

For portable instruments, a preferred source is the rechargeable Ni-Cd battery. For a ripple-free voltage reference, one chooses a mercury battery [Hg on Fig. P5]. The output voltage per cell is precisely known (e.g., 1.3524 V), and is stable to better than 0.1% during all but the last 10% of the battery's life. CMOS and op-amp circuits can be powered with the so-called transistor battery, which is usually C-Zn but with improved voltage stability and energy density. Newer battery types are emerging. For low power electronics, look for Li- and also Ag-based cells to replace the "transistor" battery. For higher current situations, look for the gel battery to replace the Ni-Cd, since the Ni-Cd tends to fail unpredictably in ordinary use. The gel battery is functionally like a car battery: rechargeable, with many amp current capability, except that the liquid electrolyte is imbedded in a gelatin-like substance.

A special family of IC voltage reference devices is now replacing the mercury battery and Zener diode in ultrastable situations. One type, available from manufacturers such as Analog Devices and National Semiconductor, is a two-terminal device functionally equivalent to a Zener diode. Another type is a three terminal device, with two power supply pins and an output terminal. Voltage accuracy is about 0.1%, and temperature coefficients are less than 10 ppm/$^{\circ}$C. Both types are called band gap reference devices: the voltage value is tied to a fundamental quantity, the semiconductor band gap, rather than to a variable such as the avalanche breakdown voltage in a conventional Zener diode.

Selecting Instruments for the Electronics Lab

Some acquaintance with basic test instruments is necessary in order to debug IC circuits. The introduction can be gradual, since the book is arranged with digital circuits first. Very little instrumentation is required to see if a signal level is high or low. By the time the reader gets to the analog portion, it is necessary to be familiar with the entire list below. A typical lab bench setup with these instruments is shown in Fig. P6.

Power supply. A basic 5 V supply is essential for TTL digital circuits. A ± 15 volt supply is needed for op amp circuits and will be useful for CMOS digital circuits as well. These should be well regulated (stable, with low ripple), and, if possible, include current-limit protection so an accidental shorting out will not be destructive.

Digital Multimeter. The cost of 3 digit digital multimeters has fallen to the point that most labs will have them, in addition to the less accurate multimeter or VOM.

Logic Probe. This device is more convenient than a meter for digital circuits, since the output is a visible on/off indication of the binary state at the point being measured. Some logic probes can only read voltage levels, while others have pulse-grabbing features to detect a fast pulse.

Function Generator. The lab bench needs a variable frequency square wave to serve as a clock for digital circuits, and a variable frequency sine wave for analog circuits. It is convenient to have both in one box, the function generator, which should be voltage-controlled if possible.

Frequency Counter. If possible, have a frequency counter available in the lab for the precise measurement of signal frequency and event timing.

Oscilloscope. This is the most difficult of the basic instruments to use. Yet, if one were to choose one instrument to take to a desert island, it would be an oscilloscope. The scope allows you to measure voltage and time, even if you don't have a digital multimeter or a frequency counter. The resolution is limited to the 1% accuracy of reading the beam's location on the scope face. The details of events are often understood only when one sees the complete waveform on a scope; a voltmeter and counter can't do that. In fact, if any IC is behaving strangely, look at it on the scope to see if the circuit is oscillating.

Graphic Recorder. If you already have all of the above, the next instrument might be the graphic recorder. It gives a more precise permanent record than a scope photo, though it is limited to much slower events (about 1 s to move across a page). The XY recorder is preferable to the strip-chart recorder, which can only record events as a function of time. The graphic recorder can aid in automating measurements. For example, the frequency dependence of a filter's response can be recorded graphically without writing down a single data point.

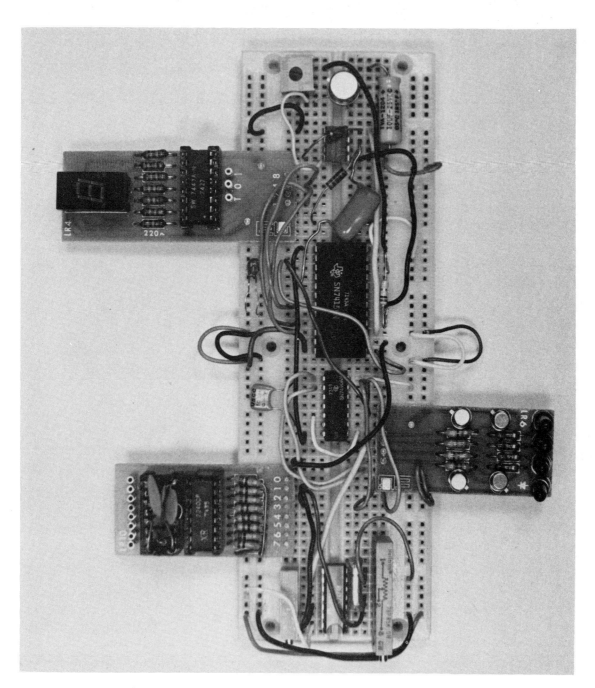

Fig. P1 A typical solderless breadboard. Active components such as transistors and integrated circuits plug in along the center line. Mini-sockets which run out to the edges bring out parallel connection points. A wire or passive component lead may be pushed into a socket, making a presure contact for quick yet adequate connections. There are <u>bus</u> <u>strips</u> runnning along the edges for power-line connections.

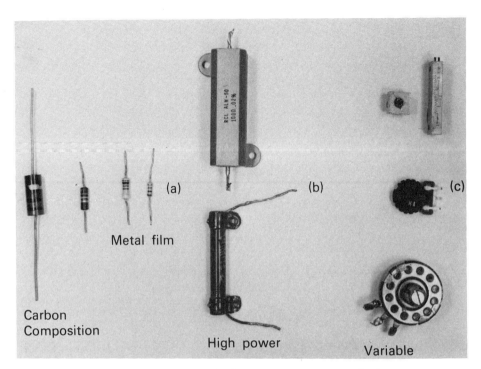

Fig. P2 Resistors: (a) Low power; (b) Higher power; (c) Trimpot or variable resistor.

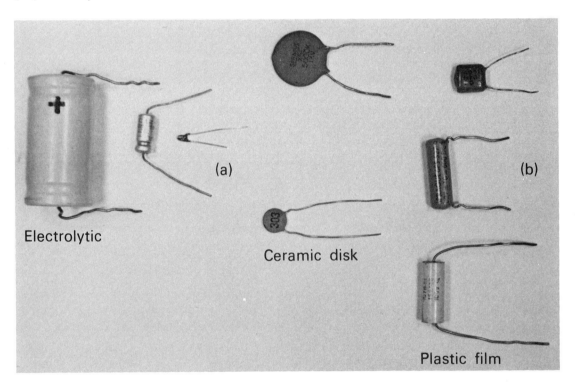

Fig. P3 Capacitors: (a) Electrolytics for power supplies (left) and decoupling (right); (b) Ceramic disk and plastic film.

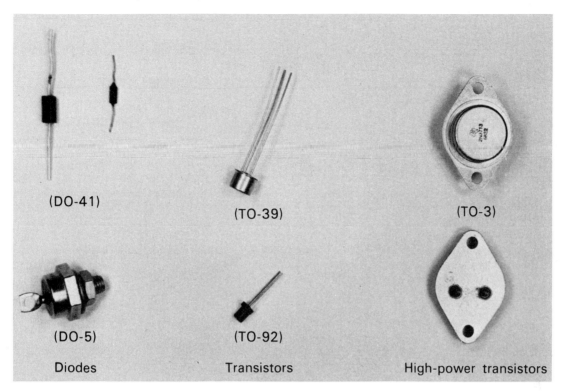

Fig. P4 Diode and transistor examples, both low and high power.

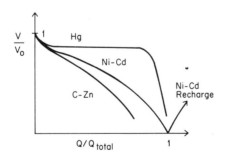

Fig. P5 Life-cycle curves for several battery types.

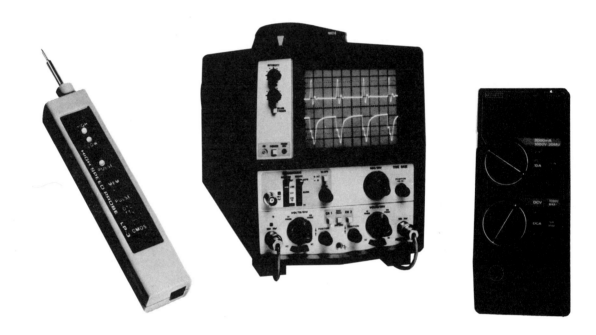

Fig. P6 Basic equipment for the electronics lab. (a) Logic probe. (b) Oscilloscope. (c) Digital multimeter.

Experiments
with
Integrated Circuits

EXPERIMENT 1:

DISCRETE COMPONENT ELECTRONIC CIRCUITS AND MEASUREMENTS

A. BACKGROUND

This experiment is remedial, for those who have never had the chance to get their hands on common electronic measuring equipment, or to build common dc and ac passive and active circuits. By the end of the experiment, you should feel comfortable with the basic equipment listed below, and will have had a chance to make the most common errors in circuit connections. Bon voyage.

B. REFERENCES
1. Text, Chapter 1.3, and especially Fig. 1.5 (a)-(d).
2. Text, Appendix A2, and especially Fig A2.4.

C. PROBLEMS
Text, Prob. 1.1 - 1.3, 1.5 and 1.15.

D. EQUIPMENT
Black boxes (2).
Volt-ohm meter (VOM).
Digital voltmeter or **digital multimeter (DMM).**
Resistors (10 K to 1 M) and **capacitors** (0.001 to 1 uf).
Oscilloscope.
Oscillator, audio frequency.
Transistor, medium power (eg., 2N3521).
LED, resistors, and battery (see Fig. 1.2).
Transistor amplifiers, common emitter and common collector (see Fig 1.3).
(Optional) **Transistor curvetracer.**
(Optional) **Differential oscilloscope.**

E. MEASUREMENTS

1. Equivalent circuits and dc measurements. This section is intended to familiarize you with the idea that a box whose insides are unknown can be characterized or modeled by an equivalent circuit which adequately describes its properties seen from the outside world. Specifically, you will learn about Thevenin's Equivalent Circuit, and the idea of circuit loading. Additional instrumentation techniques you will learn: comparing an analog multimeter (volt-ohm meter) with a digital multimeter, and a first exposure to the oscilloscope as a measuring tool (Reference 1).

    a. Box A contains an unknown voltage source. Measure it with both the multimeter and the digital voltmeter (DVM). Are there any sizeable differences? How accurately and how precisely can you tell.

    b. Box B contains an unknown voltage source, plus an unknown array of resistors. Measure the output voltage with both the multimeter and the digital voltmeter. Are there any sizeable differences? Formulate a model (using Thevenin's theorem) of what is inside that will explain the origin of the differences. Then design a test of your model that will characterize the values of all parameters of the equivalent circuit.

Concepts:
    Open-circuit voltage (How?);
    Short circuit current;
    Input impedance of meters (consult your instructor for values).

Open the box and draw the circuit inside. Can you see how the actual circuit relates to the equivalent circuit?

2. AC circuits, filters and the oscilloscope. The passive filter is among the simplest of black boxes whose input/output response or transfer function is useful in signal processing. This section is intended to introduce you to working with capacitors and resistors wired as low-pass and high-pass filters. It is also intended as an introduction to oscilloscope use (Reference 2).

a. Hook up as a low-pass filter an RC combination [text, Fig 1.8(a)] whose time constant is about 10^{-3} s. Connect a sine wave generator to the input and an oscilloscope to the output. Vary the frequency over a wide range and note qualitatively what happens to the amplitude on the screen. Find the frequency at which the output is reduced to about $(1/2)^{1/2} = 0.707$ of the input amplitude. Over what frequency range does the output amplitude fall off as $1/f$? Over what frequency range is the output a constant? Now take some numerical measurements at frequency intervals of about three points per decade above and below the cutoff frequency, and plot the transfer function.

Warning: Sine wave generators have a source impedance and can get loaded by the filter. Think of a way to measure the sine wave generator's output impedance (use a resistance box), and either make sure that the impedance level of your filter is always ten times larger, or be prepared to actually <u>measure</u> the input amplitude presented to the filter at each frequency.

b. Repeat the above with the same RC components hooked up as a high-pass filter [text, Fig. 1.8(c)]. Again explore the response qualitatively before taking any numerical data points.

3. Transistors. This section is intended to familiarize you with handling a junction or bipolar transistor. First, you will hook it up as a switch suitable for driving an LED. Then, (with a prewired circuit) you will measure the key characteristics (and differences) of the two most common transistor configurations: the common emitter amplifier, and the common collector or emitter-follower amplifier.

a. How to check if a transistor is good or not? See if both the base-emitter and the collector-base lead pairs act as diodes. This can be done with a Volt-Ohm meter measuring resistance, since a diode has a high back resistance and a low forward resistance. Try this test: what value of R is observed with the meter connected first one way, and then reversed?

This test makes it possible to determine which is the base lead on a bipolar transistor. How? Note that if the **polarity** of the internal battery in the ohm-meter is known (a second volt meter will tell you), this test can also be used to determine if an unknown transistor is **pnp** or **npn**. How?

(Optional): connect your transistor to a curve tracer and look at the characteristic curves. What is the current gain?

b. A transistor may be used as a switch or valve to control a flow of current using a much smaller current. Hook up the circuit shown in text, Fig. 1.18(c). The load is a light-emitting diode (LED) with a current-limiting resistor. The input is a battery and resistor combination selected to set a particular base current (how big?). Meters may be inserted in both the collector lead and the base lead to measure the currents there. First leave the base circuit open. What collector current flows? Is the LED on? Now complete the circuit to the base. How much current flows in the collector lead? Does the LED light? Measure the current flowing in the base lead, and calculate the current gain of the circuit.

c. Two amplifier circuits are connected as demonstrations of the two most common configurations. A common emitter configuration gives large voltage gain. A common collector configuration gives only a unity voltage gain, but a large current gain (current booster). Connect the oscillator as the input signal. Use a __differential__ oscilloscope (two probes measuring the voltage difference between two points)* to observe the input voltage and output voltage of both amplifiers. Which one inverts the sign of

---

* If unavailable, you will need a VOM or DMM with an __AC__ millamps scale. If you use a VOM, it's recommended that you set the oscillator at the line frequency.

DISCRETE COMPONENT CIRCUITS AND MEASUREMENTS   1-4

the signal? Estimate the voltage gain (to 10%). The circuit also contains small current sampling resistors. The differential voltage across the resistor is a measure of the current flow in that part of the circuit. Measure input and output current in both amplifiers, and compute the current gain of both. Which amplifier has the largest **current gain**? Which amplifier has the largest **power gain**?

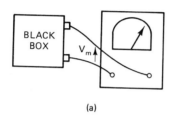 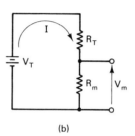

Figure 1.1 Thevenin and Norton equivalent circuits.
(a) "Black box" whose voltage is being measured.
(b) Voltage divider circuit.

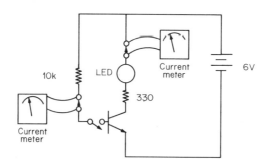

Fig. 1.2 Transistor connected as a switch to drive a light-emitting diode (LED). The resistor in series with the LED determines the current and hence the brightness. The resistor in series with the base sets the base current at a value sufficient to turn the transistor fully on when the switch is closed. Current meters can be inserted by opening the circuit at the points shown.

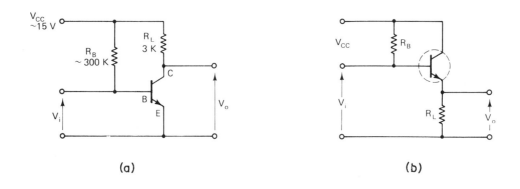

Fig. 1.3 Transistor amplifiers: (a) Common-emitter; (b) Common-collector, or emitter follower.

EXPERIMENT 2: COUNT BY TWO'S; BINARY NUMBERS

A. BACKGROUND

This experiment introduces you to the hardware we will be using and to the basic notions (software?) of digital electronics. You will learn about binary representations, and will build an IC counter with which both binary and digital displays can be compared. The strategy is to jump right in without worrying about detailed circuit operation; that comes later. For example, you will use a 7490 counter and a 7447 decoder without any prior background on what they can do or how they do it. Rather specific instructions are given about wiring the IC's and test circuits. You may find it instructive to look up IC pin diagrams in the appendix or a data book, but don't at the moment expect to understand all of the terminology or what all of the pins do.

These initial experiments are written at a rather low level, and are often repetitive. In order to jump in without knowing all details, a cookbook-like recipe is adopted at first. The recipe need not be mastered in one exposure, but sinks in gradually when used frequently.

Conventions about the base of a given number:

n_{10} is in decimal, e.g., $7_{10} = 7$

n_2 is in binary, e.g., $10_2 = 2_{10}$

n_8 is in octal, a form of binary shorthand which groups a long binary number into subgroups of 3 bits each.

If the subscript is left off, it usually (but not always!) means a decimal. When the term _digit_ is used, it means a decimal number (for those of us with 10 fingers).

B. REFERENCES

Text, Ch. 2. If more detailed help is needed, almost any digital text covers binary numbers. For further information on outboards, see Technibook I by P. R. Rony.

C. PROBLEMS

Text, problems 2.2, 2.4, 2.5, 2.6

D. EQUIPMENT

Breadboard Socket (E&L Instruments SK-10 or equivalent)
5 V power supply
Test Modules or **"Outboards"**:
    logic switches
    binary LED monitors
    dual pulser (uses a 7400 to "debounce";
        more about that later)
    clock (uses 555 timer)
    Decimal 7-segment LED display
        (uses 7447 BCD-to-7-segment decoder)
IC'S: 7490 decimal counter (BCD output)
      (optional: 7493 binary counter)
(Optional) Frequency counter

E. MEASUREMENTS

Warning: Always turn off the circuit power when inserting IC's or outboards, lest components be destroyed. The power should be left off when circuits are wired or major wiring changes made. See Guidelines for IC Nondestruction.

Another rule: never connect two TTL outputs together; the circuit won't work. This rule includes all outboard outputs.

1. Explore the electronic representation of binary 1 and 0 states, using light-emitting diodes (LED's) to display the electronic voltage levels. Plug in the LED outboard. This and all other outboards make power connections to the outer bus strips of the SK-10 breadboard. To avoid destruc-

tion of the IC's, it is essential that power connections to the bus strips are not backwards. Follow the convention that the outer bus is + 5 V. Review Fig. i. in the Lab Introduction for details of the SK-10 breadboard.

Using a wire jumper plugged into one of the LED inputs, touch the other end of the wire to the 5 V and then to the ground bus. Which one makes the LED light up? Suppose a light ON represents a logic-1 state in binary arithmetic (truth = light). Then which voltage level corresponds to a logic 1? This correspondence between logic and electronic voltage levels is sometimes called positive logic.

How could you use such a LED display to determine the logic state at some point in a circuit?

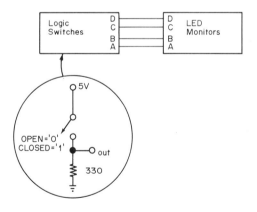

Figure 2.1 (a) The state of logic switches, a logic 1 (high) or logic 0 (low), can be tested with LED monitors. The LED's are driven by a transistor switch like Fig. 1.2.

BINARY NUMBERS   2-3

2. Explore the generation of 4-bit binary numbers using logic switches. Plug in the switch outboard. Wire jumpers across to the LED outboard as shown in Fig. 2.1(a). The actual layout is shown in Fig. 2.1(b). Fig. 2.1(a) is called a schematic diagram. The actual layout may be topologically distorted from the schematic layout. Fig. 2.1(b) is called a wiring diagram. It shows the physical details. In complicated circuits, the concepts are more clearly shown in a schematic diagram.

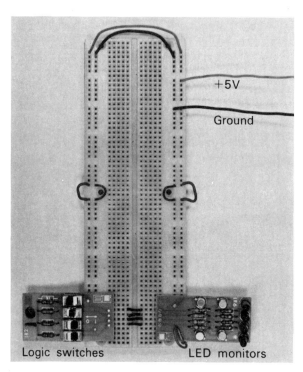

Figure 2.1 (b) Wiring diagram for part 2. The physical logic switches and LED monitors may be either individual plug-in test modules as shown (outboards), a combined digital logic plug-in breadboarding module, or functions on a separate digital logic test instrument. Plug-in modules take their power from the bus strips on the breadboard socket, and connect logic test points to rows on the socket.

2-4   BINARY NUMBERS

Begin with all switches set so no lights are lit. Verify that each switch can light a LED when its state is changed. How would you label the switch positions as <u>logic states</u> to represent what a switch does to a light?

The four switches and four lights are sufficient to represent any 4-bit binary number. Try counting in binary. Make a <u>truth table</u> showing the sequence of binary states of the logic switches corresponding to all possible switch combinations, numbered from decimal 0 through 15. Set some of these switch combinations, and verify that the corresponding binary light pattern is observed. (If one of your wires is crossed, the correspondence will not sequence in order.)

3. Using the seven-segment display outboard (decimal LED outboard), the correspondence between binary and decimal representations can be seen visually. Plug in the 7-segment display outboard, and wire it in parallel with the binary LED outboard as shown in Fig. 2.2(a). With all switches in the OFF position, turn on the power. What shows in either display outboard? Now put switch A into the ON (1) position. Do you see a 1 in the decimal display? (If not, turn off power and check your wiring.) Cycle each switch combination and record the binary and decimal numbers observed. Make a truth table with the following headings:

<u>Binary</u> <u>number</u>	Observed <u>Binary</u> <u>LED</u>	Observed <u>Decimal</u> <u>LED</u>	Expected <u>Decimal</u>
0000 to 1111			

Fill in the truth table. What do you observe on the decimal display for binary switch settings larger than 1001?

BINARY NUMBERS   2-5

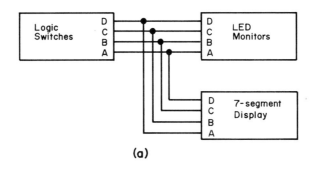

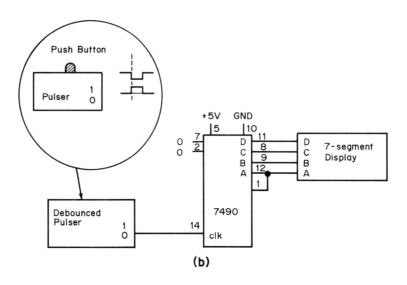

Figure 2.2 (a) A 7-segment display wired in parallel with the binary LED monitors. (b) A debounced pulser provides input pulses to increment the 7490 counter. The pulser circuit (to be studied in Chapter 4) provides clean logic transitions when the button is pushed. A <u>conceptual</u> counter wiring diagram is given. Consult the pin diagrams (Appendix X) for actual pin layout.

This decimal LED display is often called a <u>seven-segment display</u> because it represents a decimal digit with seven or less line segments. How many different symbols are possible with a total of seven segments? Since the number of possibilities with four binary inputs is larger than 10_{10}, the extras can be used as desired. Sometimes these generate 7-segment representations of the letters A-F, so the display can represent a full <u>hexadecimal</u> digit. Is your display this kind? How much of the English alphabet could you generate with these seven segments? (Although the number of segments is adequate for the number of letters, the arrangement is not.) How many binary inputs would it take to make a display capable of displaying the English alphabet? Such <u>alphanumeric</u> displays use a dot matrix, and are available not only for the English alphabet, but for others as well, such as Russian, Arabic, Japanese Kana, and even Chinese characters.

4. Add a pulser to introduce regular <u>clock</u> pulses. Plug in the pulser outboard. Wire one of its outputs to a binary LED. There are two kinds of pulser outputs, labeled 1 and 0. For the 0 output, what is the correspondence between the pulser's pushbutton switch position (out = OFF, in = ON) and LED logic state? Repeat for the pulser 1 output. The 0 output is called a <u>positive pulse</u>: low when resting, high when pulsed. Both outputs are useful, depending upon the logic level needed to drive a given circuit. Circuits with two opposite logic-state outputs occur often. The two outputs are called the <u>complements</u> of one another.

5. Add a decimal counter to be driven by the pulser. Plug in a 7490 <u>decade counter</u> IC. (Did you remember to turn off the power while doing so?) Wire the binary output pins (pins 8,9,11,12) to the decimal display outboard as shown in Fig. 2.2(b). Include the binary LED outboard in parallel. Note that the binary output pins of the counter are not lined up in sequence on the IC, and must be wired as shown to get the correct <u>weighting</u> (2^0 = A.... 2^3 = D) and <u>sequence</u> (0000......1111). A wiring diagram with suggested layout is shown in Fig. 2.3. Extra connections in Fig. 2.2(b) must be made for proper operation. Pins 2 and 7 must

be wired to ground. Momentarily disconnecting either from ground allows that pin to float high and to reset the counter to zero. Pins 1 and 12 must be wired or jumpered together. The 7490 is really two counters in one IC, and the jumper connects the two together. Other unused pins also have functions which will be explored later (see a TTL databook for details).

Cycle the pulser several times. What happens to the output display lights? Repeatedly push the pulser switch until the pattern repeats, and record the truth table for the entire cycle. (If it doesn't look like simple decimal counting, check your wiring.) Verify that you can reset the counter (pin 2 removed from ground). What happens when you cycle the pulser after a reset? Does the truth table contain all possible 4-bit combinations?

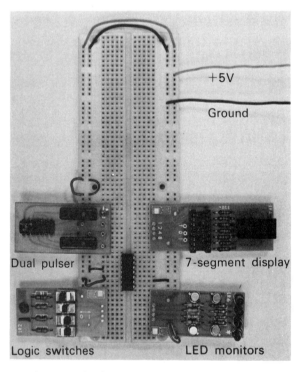

Figure 2.3 Partial wiring layout for part 5. Some of the wires have been left off for clarity. The counter IC fits neatly between the test modules.

(Optional) Replace the 7490 with a 7493 binary counter and repeat the above steps. Record the truth table over a complete counter cycle. How do the 7490 and 7493 differ? (The wiring is identical for both except that pin 7 has no function on the 7493; pin 2 is used for reset.)

(Optional) Construct an expanded truth table for the 7-segment display. For each of the 7490 counter states, record:

7490 state	Number of segments lit	Which segments lit
		a b c d e f g
0	0	0 0 0 0 0 0 0

The labels a-g denote individual display segments labeled clockwise from the top (a), with g in the center. Why are no unusual segments patterns observed? The number of segments and their location varies greatly through the cycle. The translation of a given binary input to the

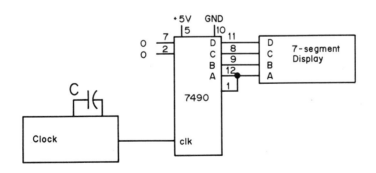

Figure 2.4 The 7490 counter may be driven by a variable frequency pulse-generator or clock, whose frequency is adjusted by the value of a capacitor and/or variable resistor.

pattern of a readable decimal number is done by a <u>decoder</u> integrated circuit, which stores a permanent representation of the correspondence. Decoders will be studied in Experiment 5.

6. Add a variable-frequency clock to drive the counter. Plug in the clock outboard. The most common clock uses an IC called the 555 timer, whose period can be varied over the range from microseconds to hours, set by an RC combination. In a typical clock module for our experiments, C of about 1 uF gives a period of about a second. Wire the circuit shown in Fig. 2.4. Using an external 1 uF capacitor, <u>calibrate</u> the clock. Measure how many pulses occur in an interval long enough to calibrate the clock to 1%. The counter can lessen the number of pulses <u>you</u> have to keep track of by a factor of 10. Now explore the variation of frequency with capacitance. Obtain several capacitors covering 1 or more decades. (Precision capacitors will give the best results.) Measure the frequency as a function of external capacitance.*

A stopwatch is adequate for periods of 0.1 s or greater, if you let the counter help you. Plot the results, and from the <u>slope</u> of the curve, determine the calibration constant of the clock. Is the deviation from a straight line adequately explained by the precision (or lack of it) of the capacitors? This calibrated clock can now be used as a high-precision capacitance meter.

(Optional) Given the above results, what C value would be needed for a frequency of 10 Hz? How about 1 KHz? Try it. Use an oscilloscope or frequency counter to measure the frequency.

---

* NOTE: Avoid electrolytic capacitors in calibrating the counter. The capacitance values of electrolytics are very imprecise, due to the changeable nature of their dielectric layer. You may wish to check this statement after doing the calibration.

(Optional) Plot a graph of frequency as a function of capacitance for frequencies faster than 10 Hz. The graph will not pass exactly through zero. Measure the free-running frequency of the outboard (no external capacitance) using a scope or frequency counter. Does it agree with the zero offset of your graph?

(Optional) Check the precision of typical capacitors. Take 10 "identical" capacitors and measure the clock frequency using each one in turn. Plot the number of capacitors with C-values in a given range (try intervals of 2%). This histogram's width measures the standard deviation of capacitance value. How does the width you obtain compare with the rated tolerance of the capacitors used?

EXPERIMENT 3: GATING CIRCUITS

A. BACKGROUND

In this experiment you will explore the input/output relationships of various kinds of gates. A gate makes it possible to start or stop a counter on command, for timing an event. In this application, one gate "enables" or "disables" the propagation of clock pulsers to a counter. In other applications, gate inputs have equal status, and are used for electronic realization of logical statements. In Exp. 3, basic gates (AND, NAND, OR, NOR) will be explored, verifying the truth tables. A few more amusing logic applications follow, explored in this experiment via truth tables.

B. REFERENCES

Text, Ch. 3. See also Lancaster, TTL Cookbook, Chapter 3, and if more help is needed, any standard digital logic text.

C. PROBLEMS

Text, Prob. 3.4 and 3.5. (Optional: Prob. 3.9). For Prob. 3.4, sketch your answers in the lab book Fig. 3.1.

D. EQUIPMENT

SK-10 socket

Outboards: LED display, pulser, clock, 7-segment display and binary display.

IC's: 7400 (NAND), 7402 (NOR), 7408 (AND), 7432 (OR)

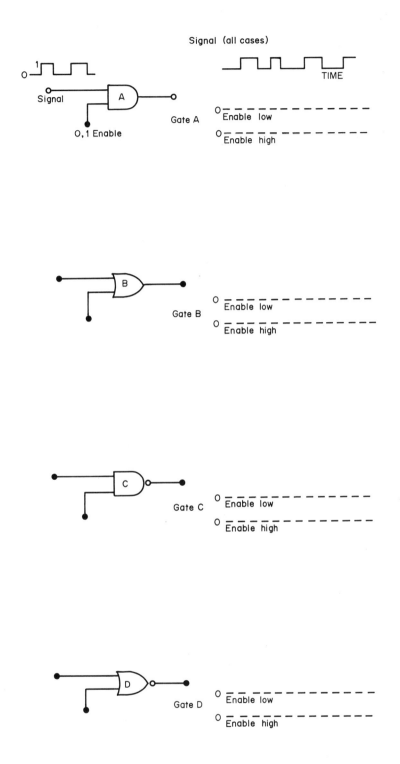

Figure 3.1 Comparison of gate functions, when one gate input <u>enables</u> the flow of data.

3-2   GATING CIRCUITS

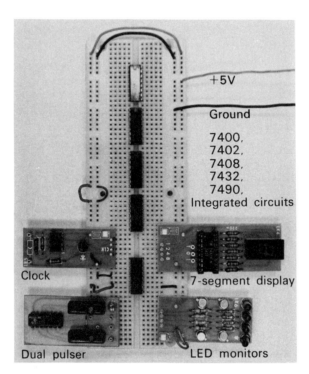

Figure 3.2 Breadboard layout for quickly comparing gate functions.

GATING CIRCUITS   3-3

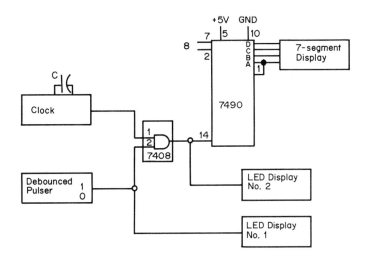

Figure 3.3 Circuit for gate testing, with 7408 AND shown as an example.

E. **MEASUREMENTS**

To avoid repetitive boredom, plug in all four IC's and the outboards needed in the configuration shown in Fig. 3.2. This will facilitate rapidly working through the operations below which compare the operation of AND, OR, NAND, and NOR, so you can get on to more interesting parts at the end of the experiment. Most TTL gates have a common convention for power pins. However, with power on diagonal corners, a gate will be destroyed if plugged in backwards. The pin convention for logic inputs and outputs varies from gate to gate, so gate IC packages cannot be interchanged without consulting the pin diagrams. Wire the circuit shown in Fig. 3.3. This allows a pulse train generated by the clock to drive a counter when the gate allows it. LED binary readouts allow a comparison between the gate output and the input which functions to <u>enable</u> or <u>disable</u> the clock pulses from reaching the counter. Note that the pulser output used is the 1 output (normally high or "true"). <u>The identical circuit will be used in Sections 1-4 below and the figure will not be repeated.</u> Merely replace the 7408 with the appropriate signal connections to another gate on the breadboard.

1. **The AND function.** Using the 7408 AND gate as shown in Fig. 3.3, note if the counter is incrementing. What is the logic state at pin 2 of the gate (the <u>enable</u> input)? What is the gate output doing, as seen by the binary LED display? Is it varying in phase (high when the clock is high) or out of phase with the clock? Sketch this time dependence of clock and gate output and compare with text Problem 3.4.

Now <u>disable</u> the counter by pressing the pulser switch in. What is the logic state of gate input 2? What is the logic state of the output? Why? Sketch clock and gate waveforms and compare with the homework problem.

Fill in the answers to the AND gate questions in Table 3.1.

2. **The NAND function.** Wire one of the gates of the 7400 (inputs pins 1,2; output pin 3) in place of the 7408. Is the counter incrementing? What is the logic state of LED display No. 1? Is the gate output varying in phase with the clock? How does this differ from the AND gate results?

Now disable the counter by pressing the pulser switch in. What is the logic state of the output? Why? Is the counter counting? How does this differ form the AND situation?

Sketch the time dependences of inputs and outputs for both pulser states and compare with homework problem 3.4. Fill in the answers to the NAND gate questions in Table 3.1.

3. **The OR function.** Replace the 7400 with one of the gates of a 7432 (pins 1,2 input, pin 3 output). Is the counter incrementing? What is the logic state of both binary LED displays? Explain. How is this different from AND or NAND? Now press the pulser switch in and note what happens. Is the counter incrementing? Record the logic states of the binary readout lights.

GATING CIRCUITS

Sketch the time dependences of input and output for both pulser states and compare with homework problem 1. Fill in the answers to the OR gate questions in Table 1.

4. **The NOR function.** Replace the 7432 with a 7402 (pins 1,2,3, but note that the logic is different). Is the counter incrementing? What is the logic state of both binary LED displays? Explain. In what ways is this similar to the OR? In what ways is it different? Now press the pulser switch in and note what happens. Is the counter incrementing? Record the logic states of the binary readout lights. Is the counter incrementing in phase or out of phase with the clock?

Sketch the time dependences of input and output for both pulser states and compare with homework problem 3.4. Fill in the answers to the NOR gate questions in Table 3.1.

5. **Truth Tables.** Using your sketches of gate input and output waveforms obtained above, write the truth table you observe for AND, NAND, OR, and NOR gates. If the correspondence between waveforms and "truth" is not clear, repeat measurements as needed (the set of measurements above in fact included all the possible input states).

6. **Unconnected inputs.** Situations occur when some gate inputs are not used, and it is important to know what to do with them to ensure proper gate operation. Using your results above, devise an experiment to determine whether an unused gate input floats high or low. Use LED lights at the output to determine what is happening. (You may not use an LED at the unconnected _input_, since it will then not act unconnected.) Check your idea with at least two gates. The answer should be the same for _all_. Now write down rules for what to do with unused inputs of AND, OR, NAND, and NOR gates. The answer is _not_ the same for all.

7. _(Optional)_ Verify one or more of the association or (more interesting) distribution theorems of Boolean algebra given in _text_ Table 2.2. A single 7408 AND and 7432 OR has enough gates on it to wire up both sides of the equivalence at the same time. Using two binary LED lights,

the equivalence can then be rapidly listed as all possible combinations of states A,B,C are input by the logic switches. Sketch your circuit clearly. Does the truth table you observe agree with what you found in the homework problem?

(Optional) Design and test one or more gate circuits which carry out electronically a logic statement originally expressed verbally. Examples:

"You can't have your cake and eat it too."

"You can fool some of the people all of the time or all of the people some of the time, but you can't fool all of the people all of the time."

"With today's lunch special, you can get soup or a salad, plus a main dish, and either coffee or dessert."

Proceed by writing the truth table you feel carries out the verbal logic statements. Then by guessing or intuition design a gate circuit which carries out this truth table. Verify in the lab that the circuit indeed carries out the truth table.

TABLE 3.1 Comparison of Gate Operation

Assume a two input gate in all cases. Fill in a separate answer for each gate (AND, NAND, OR, NOR).

1. What gating signal (high or low) <u>enables</u>?
2. What gating signal (high or low) <u>disables</u>?
3. What outputs are possible when the gate is enabled?
4. What outputs are possible when the gate is disabled?
5. Does the gate transmit a faithful replica or does it invert?
6. Find a pair of gates which have their logic function (the AND in the case of a NAND gate) satisfied by the same inputs.
7. Find another pair of gates which also have their logic function satisfied by the same inputs.

Gate:	AND	NAND	OR	NOR
Question				
1.				
2.				
3.				
4.				
5.				

3-8   GATING CIRCUITS

EXPERIMENT 4: BOOLEAN ALGEBRA IN LOGIC DESIGN

A. BACKGROUND

In this experiment you will learn to use a few of the theorems of Boolean Algebra. The most useful of these are de Morgan's theorem plus the flexible use of negative logic. These allow, for example, building nearly any logic function with one or two kinds of gates, so one doesn't need all 4 of AND, OR, NAND, NOR. As an example, the very useful exclusive-OR function will be explored in depth. Finally, gate arrays which perform complex logical operations will be explored in an example of code conversion.

This experiment should be done by all computer science majors. Students who will do logic design should also do it. Others may wish to skip it, since other techniques such as read-only memories exist to implement an arbitrary truth table, and the decreasing cost of gates has made minimization techniques less important.

B. REFERENCES
Text, Chapter 3.

C. PROBLEMS
Text, Probs. 3.10, 3.12, 3.13, 3.14. In addition:

1. Verify the equivalences shown in Fig. 3.16, using truth tables and/or Boolean algebra. These examples are useful in synthesizing unavailable gates.

2. Answer the three questions in the text about the octal-to-binary decoder (Fig. 3.28).

D. EQUIPMENT

SK-10

Outboards: 4-light binary readout; 4-switch register; clock

IC's: 7486 exclusive OR; 7400 NAND; two 7420 4-input NAND's; three 7410 3-input NAND's; other IC's as needed.

E. MEASUREMENTS

1. Select one of the following. Using 7400 NAND's only, construct an OR gate, and verify the truth table. Alternatively, using 7402 NOR's only, construct an AND gate. Use binary switches for inputs and binary LED lights for outputs.

2. Measure the truth table for an exclusive OR gate (7486). Use logic switches as inputs, and hook both inputs and the output to three separate binary light indicators to rapidly determine the states. How does the result differ from your results (previous experiment) for the inclusive OR gate?

3. Explore the application of the exclusive OR as a controllable inverter. Replace one of the switch inputs with a slow (~1 Hz) clock. Note that the pattern of blinking readout lights may be made either in-phase or out-of-phase, depending on the state of the second gate input. Is this consistent with the truth table?

4. Construct an exclusive OR gate using one of the many possible methods using NORS or NANDS alone. For example, do it using the 4 NANDS on a single 7400 IC (see Fig. 3.22). If you do it right, the truth table you observe (switches for inputs, binary lights checking states) should be identical with that observed for the 7486. If you have trouble getting the correct response, make sure that the output of each gate is carrying out its expected Boolean function.

Optional: Try another version: the 5-gate naive version (Fig. 3.21), the NOR's only version (see Prob. 3.12), or the 3-gate minimized version (Prob. 3.13).

5. Construct a 4-bit parity checker using 3 exclusive OR gates on a single 7486 IC (Text, Fig. 3.25). Use a binary light to read the output status, and a switch register as 4-bit binary input. Verify the truth table: Output true for any odd number of inputs true. If you have trouble, check the status of intermediate gates.

6. Do _one_ of the following: Construct and test a binary-to-octal (Fig. 3.27) or octal-to-binary (Fig. 3.28) decoder. Use switches for inputs and lights for outputs. Some multiple-input gates not part of your basic kit may be required.

Octal _to_ _Binary_. You will need 2 7420 IC's, which are NANDS. Reverse the input logic polarity using inverters (7404), or, even simpler, let the selected octal number go to zero while the nonselected one stays high (negative logic).

Binary _to_ _Octal_. You will need 3 7410 IC's, and two 4-bit light displays.

EXPERIMENT 5:

DIGITAL DECODING, MULTIPLEXING, AND SEQUENCING

A. BACKGROUND

Using gate arrays similar to those used in Exp. 4, it is possible to routinely implement a variety of decoding or code conversion problems. The most standard applications have been implemented on a single decoder IC. Examples include the 7442 (BCD to decimal) and the 74154 (binary to decimal). Another example is the 7447, which decodes a binary number into the correct array to light up a numeral on a 7-segment display. If the binary inputs of a 7442 or 74154 are sequentially addressed by the output of a binary counter, the IC can be used as sequencer, to sequentially select or enable a given line. Decoders allow a computer to route instructions on a common bus to one device which is addressed or selected. If a data input line also is present, that same IC can be used as a demultiplexer, steering serial binary information on a single wire to any number of outputs, depending upon which one is addressed by the binary decoder input lines. Finally, the opposite operation, multiplexing, brings many binary input lines onto one under the command of binary address lines. Multiplexing is carried out by special IC's such as the 74150 (16 to 1) or 74151 (8 to 1) multiplexer. Multiplexing is used in data transmission, and in computers which are time shared or have multiple inputs.

All these operations will be demonstrated in this experiment. There are two very nearly identical options. The first, option A, uses a 74154 IC for the 4 bit/16 line operation, plus a 74150 multiplexer. The second option, with Figures labeled B, uses a 7442 IC as a decoder, demultiplexer, and sequencer with up to 10 outputs and 3 or 4 binary bits of input; a 74151 is used as a multiplexer. Option A is typical of binary operations encountered in digital information processing. Option B is rather typical of signal processing in instruments such as decimal counters, and is the faster lab experiment (3 bits instead of 4). Option A is recommended if the IC's are available, and the instructions are given with that assumption. Circuit diagrams are also given for option B. Select one or the other.

B. REFERENCES

Text, Chapter 3, Section 7

TTL Cookbook pp. 140-148

This experiment was largely inspired by the Bugbooks, (Larsen & Rony).

C. PROBLEMS

1. Count the keys on a typewriter. What is the minimum number of binary bits needed to encode the letters and symbols (a) using upper case letters only; (b) using both upper and lower case letters. The ASCII standard code uses 7 bits. How many extra characters are therefore available beyond the symbols on your typewriter?

2. Suppose a computer is to be connected to 16 output devices. Each device responds to 8-bit data words. How many wires are required to be connected to all devices in parallel? Now suppose that the devices are slow enough so that they can be serviced by the computer one at a time. Study the multiplex-demultiplex diagram of Fig. 5.4A. Show how to apply this idea to the problem. How many address bits are required? How many 74154's and 74150's are required? How many wires are required for this multiplexed connection? (The number of wires is less than 16.)

3. Prob. 3.19 and 3.20.

D. **EQUIPMENT**
   SK-10 socket and power supply
   Outboards: LED binary display; LED digital display; pulser; clock; logic switches
   ICs: 7404 or other inverter (for either option)
      7493 4-bit binary counter, 74154 4 to 16 decoder, and 74150 16 to 4 multiplexer for Option A; or
      7442 BCD to decimal decoder and 7490 BCD counter for Option B

Note: Replacing the pulser with a slow clock makes the "data" being transmitted more interesting. Also, in observing sequencer operation, your observation of the selected channel advancing will be speedier if you use a binary LED readout of 4 or 8 channels at a time (rather than one as shown).

E. **MEASUREMENTS**
   1. **Decoder**. Truth tables for the 74154 (4-bit binary in, 16 out) and 7442 (BCD to decimal) are given in Table 5.1. Note that for a given input word, at most one output line is "selected." It happens that the selected line is low and all the others are high. This choice needs to be kept in mind in the logic attached to enable a selected instrument, for example. Suggested circuit diagrams to study decoder operation are given in Fig. 5.1A and 5.1B. Wire up one or the other. As inputs, use either a binary switch register or a counter (7490 for 7442, 7493 for 74154) driven by a pulser. The 74154 is a 24 pin IC, the first encountered in these experiments. Use extra care getting it in and out of the breadboard, and be careful with power connections, since it is fairly expensive ($4.00 list, $1.50 discount). The 74154 has two extra inputs, pins 18 (data) and 19 (enable). Wire these to ground (0) for use as a decoder.

Table 5.1: Truth Tables for Decoders

7442 (BCD to dec.)

Inputs				Outputs									
D	C	B	A	0	1	2	3	4	5	6	7	8	9
0	0	0	0	0	1	1	1	1	1	1	1	1	1
0	0	0	1	1	0	1	1	1	1	1	1	1	1
0	0	1	0	1	1	0	1	1	1	1	1	1	1
0	0	1	1	1	1	1	0	1	1	1	1	1	1
0	1	0	0	1	1	1	1	0	1	1	1	1	1
0	1	0	1	1	1	1	1	1	0	1	1	1	1
0	1	1	0	1	1	1	1	1	1	0	1	1	1
0	1	1	1	1	1	1	1	1	1	1	0	1	1
1	0	0	0	1	1	1	1	1	1	1	1	0	1
1	0	0	1	1	1	1	1	1	1	1	1	1	0

74154 (4 to 6)

Inputs				Outputs															
D	C	B	A	0	1	2	3	4	5	6	7	8	9	10	11	12	13	14	15
0	0	0	0	0	1	1	1	1	1	1	1	1	1	1	1	1	1	1	1
0	0	0	1	1	0	1	1	1	1	1	1	1	1	1	1	1	1	1	1
0	0	1	0	1	1	0	1	1	1	1	1	1	1	1	1	1	1	1	1
0	0	1	1	1	1	1	0	1	1	1	1	1	1	1	1	1	1	1	1
0	1	0	0	1	1	1	1	0	1	1	1	1	1	1	1	1	1	1	1
0	1	0	1	1	1	1	1	1	0	1	1	1	1	1	1	1	1	1	1
0	1	1	0	1	1	1	1	1	1	0	1	1	1	1	1	1	1	1	1
0	1	1	1	1	1	1	1	1	1	1	0	1	1	1	1	1	1	1	1
1	0	0	0	1	1	1	1	1	1	1	1	0	1	1	1	1	1	1	1
1	0	0	1	1	1	1	1	1	1	1	1	1	0	1	1	1	1	1	1
1	0	1	0	1	1	1	1	1	1	1	1	1	1	0	1	1	1	1	1
1	0	1	1	1	1	1	1	1	1	1	1	1	1	1	0	1	1	1	1
1	1	0	0	1	1	1	1	1	1	1	1	1	1	1	1	0	1	1	1
1	1	0	1	1	1	1	1	1	1	1	1	1	1	1	1	1	0	1	1
1	1	1	0	1	1	1	1	1	1	1	1	1	1	1	1	1	1	0	1
1	1	1	1	1	1	1	1	1	1	1	1	1	1	1	1	1	1	1	0

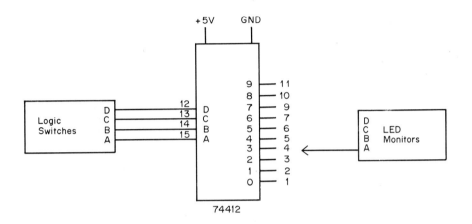

Figure 5.1A Decoder circuit diagram, using 74154.

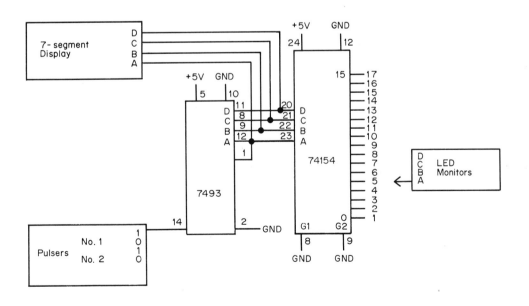

Figure 5.1B Decoder circuit diagram, using 7442.

Using binary LED readout, scan the outputs when the input word is 0000. Which output is low (selected)? Are any other outputs low? Change the input to 0001. Which output is now selected? Quickly scan through the other possible input words and verify the truth table (Table 5.1). Note for the 7442 that input words larger than 1001 result in no output being selected. Why?

2. Sequencer, with programmable number of channels. A decoder whose input word is incremented sequentially by a counter is called a sequencer. The output lines may be used to enable a sequence of devices to send data to a computer, for example. If you used a counter in the previous section, you have already seen this function. If not, replace the switch register with a counter, and use 4 or 8 binary LED's connected simultaneously to the outputs to quickly observe the sequencing of a selected channel.

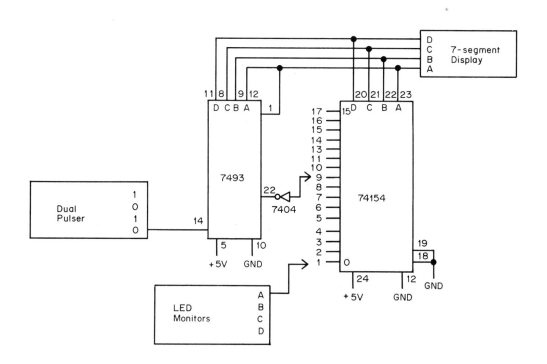

Figure 5.2(a) Sequencer circuit diagram, using 74154. The 7404 inverter input, connected to a selected 74154 output, resets the sequencer.

5-6 DIGITAL DECODING, MULTIPLEXING, AND SEQUENCING

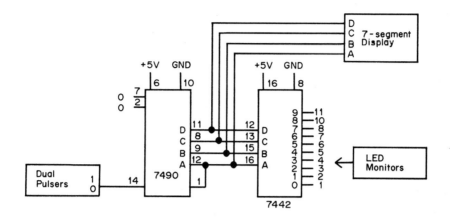

Figure 5.2B(a)   Sequencer circuit diagram, using 7442.

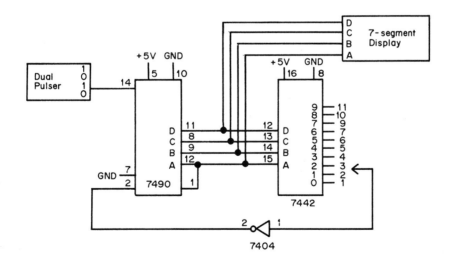

Figure 5.2B(b)   Programmable sequencer circuit diagram, using 7442.

DIGITAL DECODING, MULTIPLEXING, AND SEQUENCING   5-7

The number of channels sequenced may be determined by the sequencer output itself. This is shown in Fig. 5.2A and 5.2B. The counter reset input is wired through an inverter to one of the decoder outputs. This provides a digital "feedback" loop or programming for the number of channels to be sequenced. This kind of programming is called hard-wired (rather than by software). Looking at the circuit, suggest how you think it works. Why is the inverter needed? If the output channels are labeled 0 to n, and the reset line is wired to channel 5, what will be the highest count you will see on a 7-segment display? Now wire up the circuit and try it. Compare what you observe with what you predicted. Pay particular attention to what kind of a pulse transition the counter needs to reset. When will the reset signal occur with the circuit shown? (Don't ignore the possibility that the selected state causing the reset triggers the reset so fast that it cannot be seen on the display.) Now vary the position of the reset wire, and determine the maximum and minimum number of states which can be sequenced.

(Optional) With the counter resting in a large number state, shift the reset wire to a low number channel on the decoder. Increment the counter, and observe how the sequencer has to "wrap around" the high end of the counter's range before it gets reset.

3. Demultiplexer. With an extra data input, the decoder becomes a demultiplexer, transmitting binary data to a selected location, such as a computer output device. For the 74154, the data input (pin 18) contents (0 or 1) is transmitted to the selected output. This pin was wired low for use as a decoder. What will the truth table look like when that pin is high? The 7442 may be used as a 3-bit demultipexer (1 to 8 demultiplexer) by clever use of its truth table. Referring to table 5.1, note that if only outputs 0 - 7 are used, and only the three least significant address bits are used, the 7442 serves as a binary 1 to 8 decoder. But if the D input (pin 12) is used as a data input, the selected output will follow it high or low, just as for the 74154. Study the truth table until the meaning of this is clear.

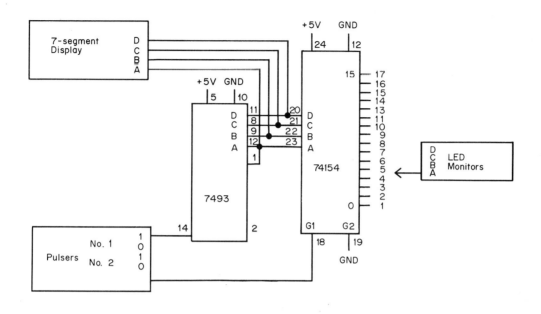

Figure 5.3A Demultiplexer circuit diagram, using 74154

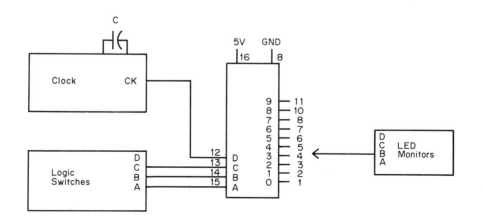

Figure 5.3B Demultiplexer circuit diagram, using 7442.

5-9 DIGITAL DECODING, MULTIPLEXING, AND SEQUENCING

Wire the demultiplexer as shown in Fig. 5.3A or 5.3B. Use a slow clock (1 Hz or so) as the "data" input. Select a given output channel with the switch register. Using an LED binary display, verify that the data stream appears at the selected output channel. What appears at the other channels? Is the output data in-phase or out-of-phase with the input data? Change the channel address, and verify that the data now appears at the new output channel.

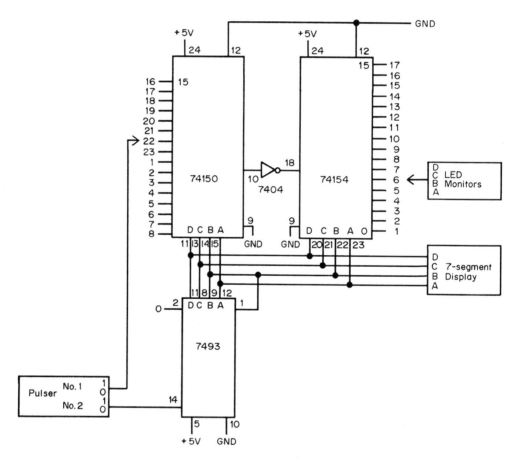

Figure 5.4A 4-bit multiplex-demultiplex circuit.

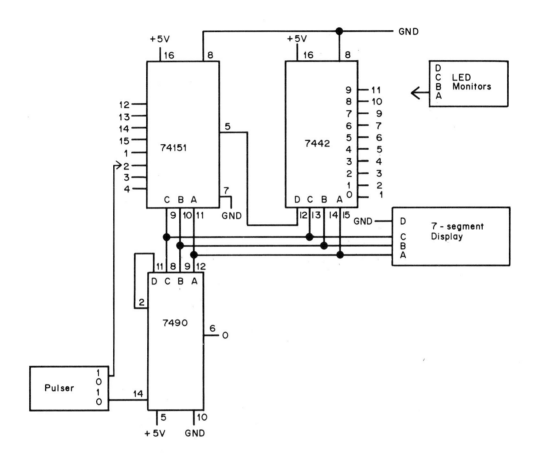

Figure 5.4B 4-bit multiplex-demultiplex circuit.

4. Multiplexing and Demultiplexing. This combination is used for data transmission over the minimum number of wires. For M devices or addresses, the number of wires is reduced from M to $\log_2(M)$. In the example we shall study, the improvement (16 to 4) is not spectacular, but consider the reduction in wires if M is 1024!

DIGITAL DECODING, MULTIPLEXING, AND SEQUENCING   5-11

Wire the circuit shown in Fig. 5.4A or 5.4B. It will all fit on one SK-10 breadboard. The inverter is required to keep the output data in phase with the input data. You may wish to replace the data source (pulser no. 2) with a clock. Increment the counter to channel 5. What pin on the 74150 will transmit data? On what pin on the 74154 will output data appear? Wire two LED's to those pins and test your idea. Does the data get faithfully transmitted? Now move the data input to another pin on the multiplexer. What appears now on the demultiplexer at the selected output? At any other output? Explain. Increment the counter until the address selected corresponds to the input you are now using for data. Does data now appear on the demultiplexer? At what output?

Why are pins 9 on the 74150 and 19 on the 74154 shown both wired to ground? In applications where the multiplexer and demultiplexer are separated by a long distance, how many total wires will be required to pass information? (Power may be provided locally, but what about a ground reference between the two devices so data levels are clearly established?)

(Optional) Replace the pulser driving the 7493 with a clock. This makes a scanner. If the clock is faster than 0.1 sec, an apparently simultaneous output display will appear on LED's connected to the demultiplexer outputs. The data displayed will be the same as at the multiplexer inputs - even though only one data wire connects multiplexer to demultiplexer. Such scanning or strobing is the basis for sharing wires in many applications, such as analog to digital conversion, calculator displays, telephone switching, and electronic music keyboards.

EXPERIMENT 6: FLIP FLOPS

A. BACKGROUND

In this experiment, you will first of all construct a basic flip flop from cross-coupled gates. Then you will explore the subtle difference between three common kinds of FF's used as latches. You will then briefly look at the truth table for the versatile J-K FF. Finally, for the brave, you will construct a clocked latch to "grab" a particular number upon command.

This experiment is significantly more difficult than the previous ones, but it is very fundamental. A mastery now of the subtle differences between different types of flip flops can save much pain later in debugging a poorly designed circuit.

B. REFERENCES

Text Chapter 4. Study especially carefully the rules for JK master-slave flip flop change-of-state (Section 4.4). See also TTL Cookbook Ch. 5.

This experiment was originally inspired by the Bugbooks, (Larsen & Rony).

PROBLEMS

1. Study the references and Fig 6.1 carefully until the answers to the following questions are clear:

a. What are the differences in the truth table of the following FF types: SR, JK, D, T?

b. What is the difference between level and edge-triggering?

c. What is a master-slave FF? Can all of the above types be master-slave?

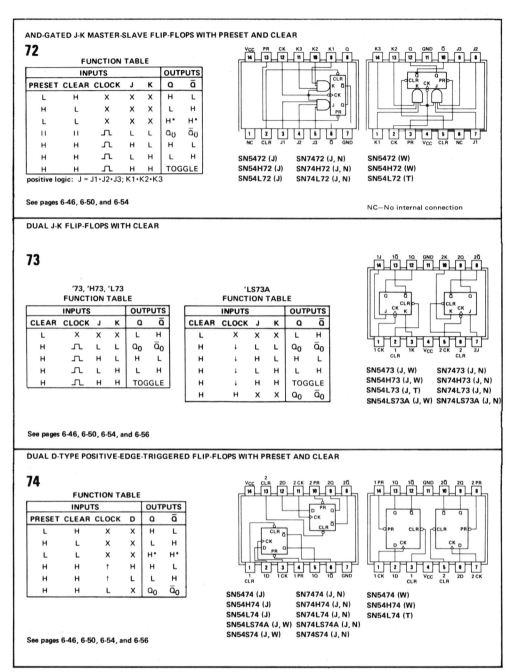

Figure 6.1 54/74 families of compatible TTL circuits.

6-2   FLIP FLOPS

4-BIT BISTABLE LATCHES

75

FUNCTION TABLE
(Each Latch)

INPUTS		OUTPUTS	
D	G	Q	$\bar{Q}$
L	H	L	H
H	H	H	L
X	L	Q_0	$\bar{Q}_0$

H = high level, L = low level, X = irrelevant
Q_0 = the level of Q before the high-to-low transistion of G

See page 7-35

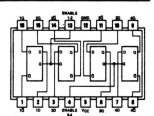

SN5475 (J, W)   SN7475 (J, N)
SN54L75 (J)   SN74L75 (J, N)
SN54LS75 (J, W)   SN74LS75 (J, N)

DUAL J-K FLIP-FLOPS WITH PRESET AND CLEAR

76

'76, 'H76
FUNCTION TABLE

INPUTS					OUTPUTS	
PRESET	CLEAR	CLOCK	J	K	Q	$\bar{Q}$
L	H	X	X	X	H	L
H	L	X	X	X	L	H
L	L	X	X	X	H*	H*
H	H	⎍	L	L	Q_0	$\bar{Q}_0$
H	H	⎍	H	L	H	L
H	H	⎍	L	H	L	H
H	H	⎍	H	H	TOGGLE	

'LS76A
FUNCTION TABLE

INPUTS					OUTPUTS	
PRESET	CLEAR	CLOCK	J	K	Q	$\bar{Q}$
L	H	X	X	X	H	L
H	L	X	X	X	L	H
L	L	X	X	X	H*	H*
H	H	↓	L	L	Q_0	$\bar{Q}_0$
H	H	↓	H	L	H	L
H	H	↓	L	H	L	H
H	H	↓	H	H	TOGGLE	
H	H	H	X	X	Q_0	$\bar{Q}_0$

See pages 6-46, 6-50, and 6-56

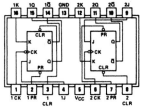

SN5476 (J, W)   SN7476 (J, N)
SN54H76 (J, W)   SN74H76 (J, N)
SN54LS76A (J, W)   SN74LS76A (J, N)

Figure 6.1 Continued.

FLIP FLOPS   6-3

    d. What is meant by <u>clocked</u> logic as applied to FF's?
    e. What is the difference between a flip flop and a latch?

  2. Text, Problems 4.3, 4.4, and 4.9.

D. EQUIPMENT
    SK-10 Socket and all outboards
    IC's: 7474 (D-type), 7475 (latch), 7476 (JK) FF's
          7400 and 7402 gates
          7404 inverters

E. MEASUREMENTS I: <u>Cross-Coupled Gates</u>

  1. Build a Flip-Flop with two inverters (1/2 7404; Fig 4.2a) "chasing each other's tail." Read the state with a binary readout light. Now see if you can drive the FF into the opposite state by forcing a high point low or a low point high, with a shorting wire to V_{CC} or to ground.

    Now put a 500 Ohm resistor in series with the ground wire and repeat the above. Can you now reliably change the state of the FF? This difference in behavior (and dependence upon detailed component values) is the reason gates are used in preference to inverters; gate inputs don't interact.

  2. Build a FF with two cross-coupled NAND gates (1/2 7400; Fig. 4.3a). Construct the truth table for this FF. What, in fact, happens when both inputs are simultaneously grounded? Somebody wins, but because the winner depends on circuit details which are not readily controllable (and vary from unit to unit) this state is not used, and is called <u>disallowed</u>.

    Now explore the operation of this FF as a contact debouncer. Change the circuit by installing pull-up resistors and a single input connection as in Fig 4.4. You may simply use a wire between ground and one FF input rather than a switch. The precise values of the pull-up resistors are unimportant (200 Ohms < R < 2K Ohms).

(Optional) Use a scope to look at the transition. Or, use a digital counter to compare the cleanness of the transition with that of a switch.

MEASUREMENTS II: IC Flip Flops
1. Comparison of latches

Wire up with parallel inputs three IC FF's as shown in Fig. 6.2.

7474 D-type (positive edge triggered)
7475 Latch (level sensing)
7476 J-K (negative level triggered). The 7476 may not work well unless the preset and clear inputs are tied to +5V when not in use.

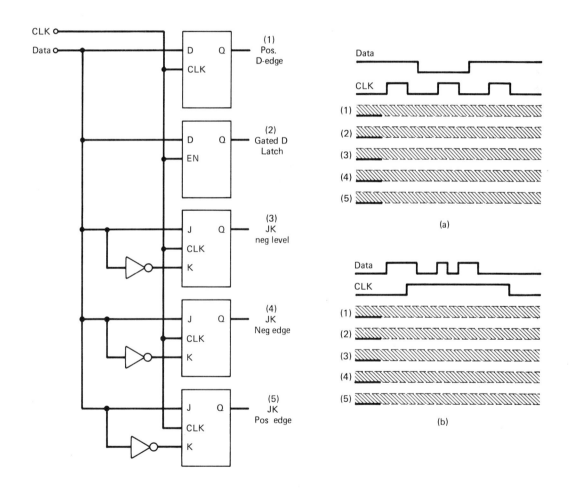

Figure 6.2 D-type flip flop circuits and waveforms.

Use LED readout lights to see the output Q of each FF. A single 7404 inverter is needed to make the 7476 into a latch. Why? Use debounced pulsers for both clock and D-inputs so one can vary independently the on-off times.

Explore and compare the response to each of the following inputs:

    a. Clock high and low. No D input action.
    b. D high and low. No Clock input action.
    c. Push Clock high. Make D high, then low. Release Clock.
    d. Push Clock high. Push D high and Hold. Release Clock, then release D.
    e. Push D high and hold. Push Clock high and release. Release D. (Between each test, initialize by cycling the clock until all lights are out.)

Based on the above, make a separate statement summarizing what it takes to reliably latch information into each of the three. Notice that the 7476 JK is "level triggered" only in a certain sense. The first high while the clock is high is what matters, but an output change only shows up when the clock goes low. This is typical of master-slave operation.

Cycle through the pattern of clock and data input combinations of Problem 4.3. Sketch the results, and compare with what you predicted. Explain any differences.

2. **Direct Set and Clear.**

Explore the set and clear inputs for the 7474 D-type and 7476 JK, using a wire to ground. Verify that these inputs take precedence over the D and C inputs.

3. **J-K Flip Flop**

Remove the inverter connection from the 7476, and determine the truth table for J and K inputs wired to 1 or 0. Note especially the Toggle connection: J and K wired to a single data input.

What function does this FF perform if J and K are held high and the clock goes through several cycles? For any input combination, when (in the clock cycle) do changes occur? Repeat the sequence shown in Fig. 6.3. The 7476 is the 3rd type shown (negative edge triggered).

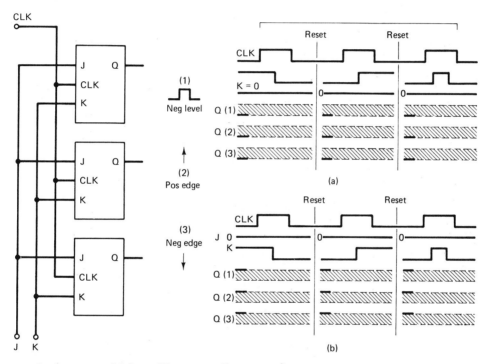

Figure 6.3 JK flip flop and waveforms.

4. Latching a Number Upon Command.   (Optional)

This experiment illustrates both the use of a 7475 quad latch, and the use of all of inputs of a 7474 D-type edge-triggered FF. Circuit operation is not easy to understand, but well worth mastering. It illustrates a common use of flip flops as "glitch detectors," detecting events at high speed. Glitches, usually undesired, are fast pulses in a circuit which arise through some subtlety of circuit

FLIP FLOPS   6-7

operation or timing. Because TTL is so fast, it can respond to a glitch as if it were an intended pulse, causing unexpected circuit response which is very hard to debug. A glitch detector holds glitches long enough so you can look at them. We will see other examples in later experiments, for example, in counters detecting the reset glitch in a divide-by-N counter, or the in-between states of a ripple counter.

Construct the circuit shown in Fig. 6.4. It will fit on one SK-10 breadboard. Consult your teaching assistant to see if a pre-built version is available to save you time. Study carefully the circuit and ponder what you think it might do. The following features are particularly important:

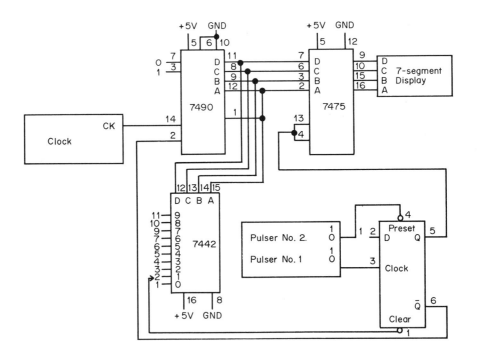

Figure 6.4 Glitch Detector circuit.

6-8   FLIP FLOPS

Pulser 1 drives the 7474 FF's clock. Since the D input is wired high, the 7474 will go high whenever it sees a clock pulse.

Pulse 2 also drives the 7474 FF, but at the PRESET input. The output will go high when the preset is high.

The 7475 latch is enabled and follows the input when pins 13 and 4 are high, and latches when they go low. That point is driven by the Q output of the 7474, which is high when either the clock or preset input drives it high.

The CLEAR input of the 7474 is driven by the selected state of the 7442. Since the CLEAR is activated by a low, the FF will clear when the selected state is reached (selected state is a 0).

The 7490 resets to zero when the RESET input goes low. This input is driven by Q of the 7575.

Given these hints, circuit operation should become clear by doing the following experiments. When power is turned on, what do you observe? Press and release pulser 1. The counter should reset, and cycle until the number selected by the 7442 (which drives the 7474's CLEAR) is reached, whereupon it will latch that number. Press pulser 2. Latching action should be inhibited, and the display will be observed to cycle through a number of states determined by the selected decoder output which drives the CLEAR.

Repeat and/or test with LED readouts various portions of the circuit until the answers to the following become clear.

Why does pulser 1 cause a cycle which stops at the selected number?

Why does pulser 2 inhibit the latching but still allow the counter to reset?

What happens to the 7490 counter after latching occurs?

What happens to the latch when the counter is reset?

FLIP FLOPS   6-9

EXPERIMENT 7: COUNTERS

A. BACKGROUND

In this experiment we will explore counters built from J-K flip flops:
  up/down;
  asynchronous (ripple)/synchronous;
  Binary/BCD

We then move to MSI counters. Taking the 7490 BCD counter as an example, we explore dividing by n using logic, dividing by 5 using the 7490's internal construction, and finally dividing by 10 with an unusual weighting by a rearrangement of wiring.

Finally, we explore a very versatile MSI counter, the 74193: it goes up/down, is synchronous, has a preset, and allows **modulo n** in a simple manner.

B. REFERENCES

Text Chapter 5. See also Lancaster, TTL Cookbook, Chapter 6

C. PROBLEMS

1. Text, Probs. 5.2, 5.3, 5.4, 5.5, and 5.6

D. EQUIPMENT
  Breadboard socket
  IC's: 7476 dual JK flip flops (two)
    7490 BCD counter
    74193 Binary up/down counter
    Gates as needed

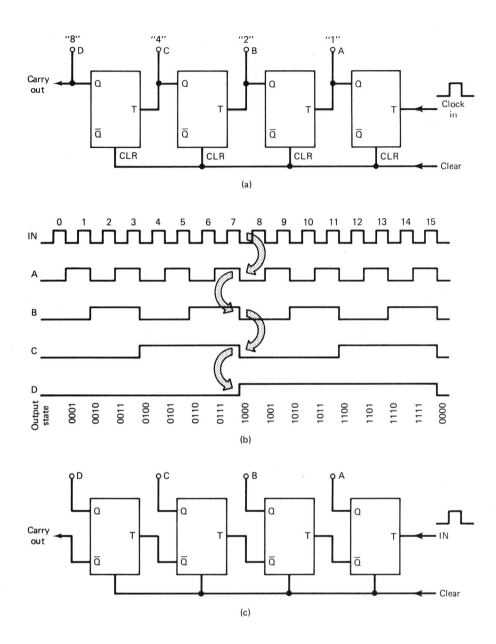

Figure 7.1 Four-bit ripple counter. (a) Circuit of up counter; (b) Waveforms of up-counter, with an example of a toggling transition rippling through; (c) Circuit of down-counter.

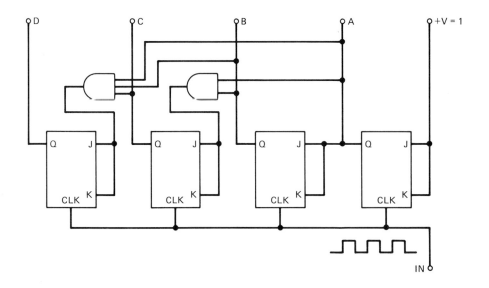

Figure 7.2 Synchronous counter circuit.

E. MEASUREMENTS I: Counters from Flip Flops

1. Construct and test a 4-bit binary asynchronous up counter (Fig. 7.1a). Use 7476 Flip Flops with the J and K input wired high. The toggle (T) input is the CLK input. Verify the waveforms of Fig. 7.1(b).

2. Change it to a down-counter (Fig. 7.1(c)) and verify that it works. What waveform pattern is observed?

3. (Optional) What simple gating scheme (Text problem 5.3) could make the up/down choice selectable by a single logic level? Try it.

4. Construct a 4-bit synchronous up counter (Fig. 7.2). If possible, leave the asynchronous version also connected. Compare the operation of the two. At low speeds, no difference will be apparent.

TTL MSI

TYPES SN5490A, SN5492A, SN5493A, SN54L90, SN54L93, SN7490A, SN7492A, SN7493A, SN74L90, SN74L93 DECADE, DIVIDE-BY-TWELVE, AND BINARY COUNTERS

BULLETIN NO. DL-S 7211807, DECEMBER 1972

'90A, 'L90 . . . DECADE COUNTERS

'92A . . . DIVIDE-BY-TWELVE COUNTER

'93A, 'L93 . . . 4-BIT BINARY COUNTERS

description

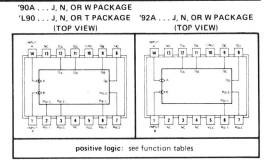

Each of these monolithic counters contains four master-slave flip-flops and additional gating to provide a divide-by-two counter and a three-stage binary counter for which the count cycle length is divide-by-five for the '90A and 'L90, divide-by-six for the '92A, and divide-by-eight for the '93A and 'L93.

All of these counters have a gated zero reset and the '90A and 'L90 also have gated set-to-nine inputs for use in BCD nine's complement applications.

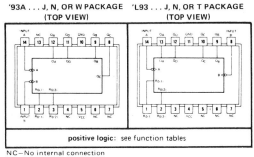

To use their maximum count length (decade, divide-by-twelve, or four-bit binary) of these counters, the B input is connected to the Q_A output. The input count pulses are applied to input A and the outputs are as described in the appropriate function table. A symmetrical divide-by-ten count can be obtained from the '90A or 'L90 counters by connecting the Q_D output to the A input and applying the input count to the B input which gives a divide-by-ten square wave at output Q_A.

NC—No internal connection

functional block diagrams

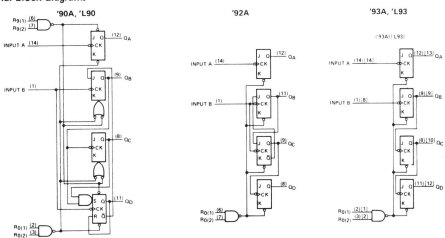

↓ dynamic input activated by transition from a high level to a low level.
The J and K inputs shown without connection are for reference only and are functionally at a high level.

Figure 7.3 Texas Instruments specifications sheets.

TYPES SN54192, SN54193, SN54L192, SN54L193, SN54LS192, SN54LS193
SN74192, SN74193, SN74L192, SN74L193, SN74LS192, SN74LS193
SYNCHRONOUS 4-BIT UP/DOWN COUNTERS (DUAL CLOCK WITH CLEAR)

BULLETIN NO. DL-S 7211828, DECEMBER 1972

- Cascading Circuitry Provided Internally
- Synchronous Operation
- Individual Preset to Each Flip-Flop
- Fully Independent Clear Input

TYPES	TYPICAL MAXIMUM COUNT FREQUENCY	TYPICAL POWER DISSIPATION
'192, '193	32 MHz	325 mW
'L192, 'L193	7 MHz	43 mW
'LS192, 'LS193	32 MHz	85 mW

'192, '193 . . . J, N, OR W PACKAGE
'L192, 'L193 . . . J OR N PACKAGE
'LS192, 'LS193 . . . J, N, OR W PACKAGE
(TOP VIEW)

logic: Low input to load sets Q_A = A, Q_B = B, Q_C = C, and Q_D = D.

description

These monolithic circuits are synchronous reversible (up/down) counters having a complexity of 55 equivalent gates. The '192, 'L192, and LS192 circuits are BCD counters and the '193, 'L193 and 'LS193 are 4-bit binary counters. Synchronous operation is provided by having all flip-flops clocked simultaneously so that the outputs change coincidently with each other when so instructed by the steering logic. This mode of operation eliminates the output counting spikes which are normally associated with asynchronous (ripple-clock) counters.

The outputs of the four master-slave flip-flops are triggered by a low-to-high-level transition of either count (clock) input. The direction of counting is determined by which count inputs is pulsed while the other count input is high.

All four counters are fully programmable; that is, each output may be preset to either desired data at the data inputs while the load input is low. The output will change to agree with the data inputs independently of the count pulses. This feature allows the counters to be used as modulo-N dividers by simply modifying the count length with the preset inputs.

A clear input has been provided which forces all outputs to the low level when a high level is applied. The clear function is independent of the count and load inputs. The clear, count, and load inputs are buffered to lower the drive requirements. This reduces the number of clock drivers, etc., required for long words.

These counters were designed to be cascaded without the need for external circuitry. Both borrow and carry outputs are available to cascade both the up- and down-counting functions. The borrow output produces a pulse equal in width to the count-down input when the counter underflows. Similarly, the carry output produces a pulse equal in width to the count-down input when an overflow condition exists. The counters can then be easily cascaded by feeding the borrow and carry outputs to the count-down and count-up inputs respectively of the succeeding counter.

functional block diagrams

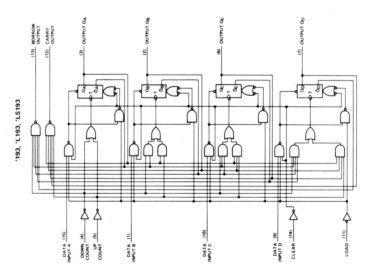

Figure 7.3 TI specifications sheets (Continued).

COUNTERS

5. Make a "glitch detector" (Fig. 7.5) which will notice the "extra states" which occur when the ripple counter (Fig. 7.1a changes state. One light monitors when the AND is satisfied. The other monitors the output of a FF which toggles whenever the selected state ($[10]_{10}$ in this example, or 1010) comes by. Do you notice any <u>extra toggling</u> during the full cycle of the counter? Explain with bit patterns during the transition. Now test with the synchronous counter. Are the same extra states observed?

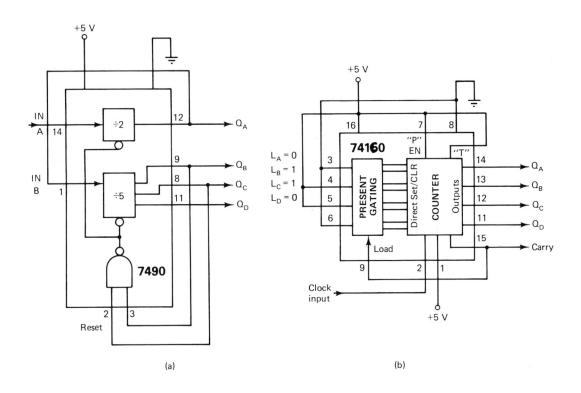

Figure 7.4 (a) Divide by 6 counter, using reset. (b) Divide by 9 counter, using preset.

MEASUREMENTS II: MSI Counters

6. You are well familiar with the 7490 BCD counter. Two new features will be explored.

a. Divide by n using the Clear. Construct and test a divide by 6 circuit (Fig 7.4a) using the appropriate outputs applied to the 0-set input. Study the waveforms and also the internal circuit (Fig. 7.3). How does it work? How would you divide by 7? Try it. (Optional) Using the glitch detector (Fig 7.5), look for a short-lived 1010 output glitch at the reset transition.

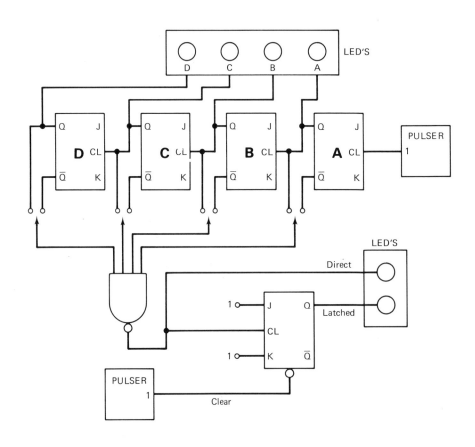

Figure 7.5 Latching an intermediate state on "glitch" of a ripple counter. The J and K inputs will float high if left unconnected as shown, but it is preferable to wire them high.

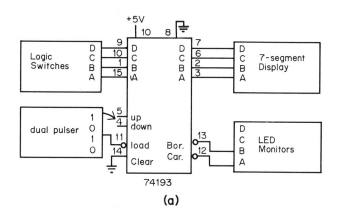

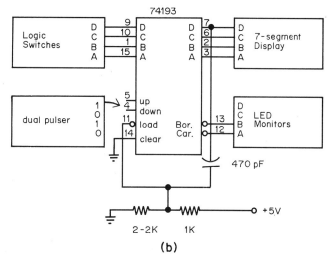

Fig. 7.6 (a) Test circuit for 4-bit synchronous counter with preset; (b) Modulo-n using Preset, with Load pulse generated by (differentiated) D output falling.

7-8   COUNTERS

b. The 7490 is actually two counters, a divide by 5 and a divide by 2 (Fig. 7.3). Explore the operation of the divide by 5 portion, graphing all waveforms over a complete cycle. Compare with text Fig 5.7.

c. A very different "weighting" is possible with the 7490, resulting in a divide by 10 output which is a symmetrical square wave. Jumper Q_8 to Input A with Input B as input. Graph all waveforms over a complete cycle. Explain the "divide by 10" symmetrical "output"(Q_1). How does the "weighting" of this counter differ from the conventional 1248 BCD?

7. Finally we deal with the 74193, the most acrobatic of 4-bit counters: it goes up or down, it is synchronous; it has both Borrow and Carry; it has an arbitrary Preset that can be used to advantage to construct a modulo n counter. Wire the 74193 (Fig. 7.6a) with logic switches driving the Preset, and pulsers driving the Clock Up/Down and preset Load. Use a 7-segment display to divide the output by 10, and lamp monitors to look at the Borrow and Carry. A Clear is provided by a wire momentarily brought to 1 but left at 0 otherwise.

Explore the Preset. Set various 4-bit combinations with logic switches, and load them with the Load pulser. Record your results.

Explore Up counting. Preset to 0111, and with the Load pulser connected to the Up clock (pin 5) verify that the counter sequences up through a full 16 number cycle.

Observe the Carry state during a cycle. Precisely when does it change from being high? Graph the waveforms of clock and carry. How could this be used in cascading several 4-bit counters into a higher modulus?

Explore Down counting. Preset to 0111, and with the Load pulser connected to the Down clock (pin 4), verify that the counter sequences down through a full 16-number cycle.

Observe the Borrow state during a cycle. Precisely when does it change from being high? Graph the waveform of clock and borrow. How could this be used in cascading a Down counter?

Now for the modulo n using Preset. Wire the circuit shown in Fig. 7.6b. The Preset **contents** is set by the switch register, but the timing of **when** the Preset occurs is set by when the most significant bit goes low. The capacitor transmits a negative-going pulse which wins over the DC high state set by the resistive voltage divider. Explore the **modulus n.** Note that n is the complement of the Preset. Why? (Text, Prob. 5.5)

EXPERIMENT 8:

DIGITAL WAVE GENERATION AND WAVESHAPING*

A. BACKGROUND

This experiment completes the basics of digital logic by introducing circuits used for timing and for restoring waveforms to clean binary form.

The range of circuits covered is wide, and numerous measurements are required to verify circuit performance. To avoid unnecessary time spent boringly, pay special attention to the hints given in selecting a minimum set of component values to adequately characterize circuit behavior.

B. REFERENCES

Text, Chapter 6.

W. C. Jung. "The IC time Machine," Popular Electronics, Nov. 73, p. 54.

SIGNETICS, Application Notes.

C. PROBLEMS

1. Text, Problems 6.1 and 6.4.

2. Pick one of the 555 applications that you want to try from the Signetics application notes. Explain what it does and how it works.

----

* A scope is used extensively in this lab. Consult your teaching assistant for an introduction if you are unfamiliar with oscilloscopes.

3. Design a digital phase shifter. The input is a sine wave. The output is a narrow pulse (sync pulse) whose phase with respect to the zero crossing of the input may be varied over 180° linearly by varying a single resistor.

D. EQUIPMENT

LED readout light, debounced pulser, seven-segment display outboards.
Oscilloscope*.
Audio Oscillator.
Multimeter or (better) **digital voltmeter.**
Pulse generator of variable width (or construct one as directed).
Frequency counter (optional but useful).
IC's: Gates as needed (7400, 7402, or 7404)
        7490 counter
        555 Timer and 74127 one-shot.

E. MEASUREMENTS

1. Schmitt Trigger with Gates. Build a Schmitt trigger with inverters (7404) or gates wired as inverters (7400 or 7402). Use component values shown in Fig. 8.1. Explore the hysteresis in the turn-on and turn-off behavior. Pick one of the following methods.

(Method one) Using a voltage divider on the 5V source, vary the input voltage slowly while observing the output using an LED readout light. Measure the turn-on voltage (to 1%) using a multimeter or DVM. Now vary the input voltage back down again. At what value does the Schmitt turn off again? How much hysteresis is observed?

(Method two) Use a sine wave oscillator in the audio range as input voltage. Two specifications are important for the oscillator. First, to insure a reliable low state of the

----

* A scope is used extensively in this lab. Consult your teaching assistant for an introduction if you are unfamiliar with oscilloscopes.

Schmitt, a low source impedance is necessary. A 50 or 100 Ohm generator is optimum. A more common 600 Ohm generator will not work directly, but if the output voltage is large enough it can be used with a voltage divider acting to lower the source impedance (Figure 8.1c). The diode shown insures that the input voltage stays positive. Some ac function generators include a dc offset control to do this. The alternative shown in Fig. 8.1(c) uses a diode in series to eliminate the negative half cycle.

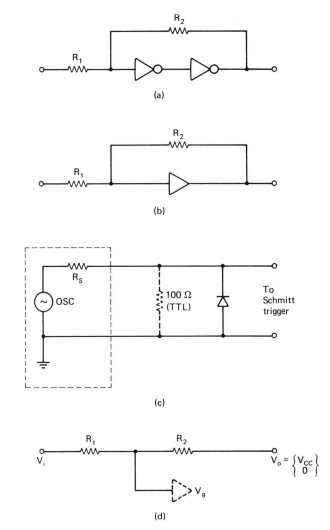

Figure 8.1 Schmitt trigger circuit with gates.

Replace the 6.8 K resistor with a 10 K pot and explore how the amount of hysteresis varies with the ratio R_2/R_1. Compare your results with the analysis outlined in the text. If you are using an oscillator as the source, this part can be done rather quickly with sufficient accuracy on an oscilloscope.

(Optional) An excellent source of dirty pulses can be made using an ordinary switch (such as the binary switch register outboard) to generate a low-high transition. Hook up a 7490 counter and 7-segment display and verify that it is nearly impossible to generate a single count with a switch transition. Now put the Schmitt trigger in between the switch and counter. Do you see single or multiple counts now? It might be useful to adjust the hysteresis range to optimize the result. Or, try a capacitor (.1 uF) on the switch output.

(Optional) What happens if R_1 becomes too large? Replace R_1 with a 1 K pot and observe the triggering and hysteresis of the Schmitt trigger.

2. Monostable Multivibrator with Gates. The procedure of this and later sections will vary depending upon whether you are using a scope and pulse generator or LED readout and debounced switch to trigger the multivibrator. In the first case, select R and C to make the pulse widths 1 msec or shorter. In the second case, select R and C to give pulses of the order of 1 sec in width. Use of a scope and short pulses is far preferable in looking at circuit operation and is also more typical of situations normally encountered. Instructions are written assuming use of a scope. A commercial pulse generator is the simplest source of pulses of variable known width. If unavailable, a quite adequate pulse source can be made from either a 555 or 74121, using a pot to vary the width. The multivibrators in this section differ in their requirements for normal input state (high or low). This will determine which output of the pulser is used, the Q or Q̄.

Construct <u>one</u> of the multivibrators from gates shown in Fig. 8.2; versions b or c are recommended.

With the output pulse width constant, vary the input pulse width, and observe what happens to the output pulse width as the input width passes through the value of the output width. Does what you see make sense in terms of circuit operation?

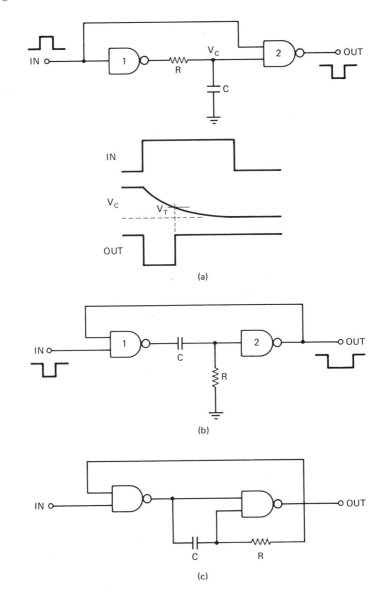

Figure 8.2 One-shot circuits constructed from gates: (a) Input pulse longer than output pulse; (b) Input pulse shorter than output pulse; (c) Another version.

WAVE GENERATION AND WAVESHAPING   8-5

Vary the R and C of your multivibrator. Select two values of R and C of your multivibrator. Select two values of R spanning half a decade, and several values of C spanning two or more decades. Measure and plot the output pulse width as a function of C and of the RC product. (Note: R should remain below 500 Ohms for reliable operation.)

(Optional) Vary R above 500 Ohms and observe when erratic or unsatisfactory operation occurs.

(Optional) Observe charging or discharging waveforms at the capacitor on an oscilloscope. The operation is clearest if a dual trace scope is used, or if the sweep is triggered from the input pulse leading edge. Do the waveforms make sense in terms of circuit operation?

3. Integrated Circuit Monostable Multivibrator. Hook up a 74121 multivibrator, using the pin diagrams and wiring information given in Fig. 8.3 as input a fixed narrow (1 us or less) pulse. For a fixed C, vary R over several decades. Three values per decade is adequate. Measure and plot as you do so the pulse width. Now repeat for a fixed R, varying C. Compare the scaling (or lack of it) with the RC product.

(Optional) For small values of external C, the circuit behavior may be a non-linear function of C, due to additional stray or internal capacitance. Vary C towards pulse widths at the lower limit of time resolution of the scope you are using, and see if the pulse width begins not to track.

(Optional) Hook up a 555 timer as a one-shot, as shown in text, Fig. 6.12. Use RC values to bring the pulse width to ~1 sec in width. What is the longest pulse you can make with components available? What happens when the input pulse becomes longer than the output pulse?

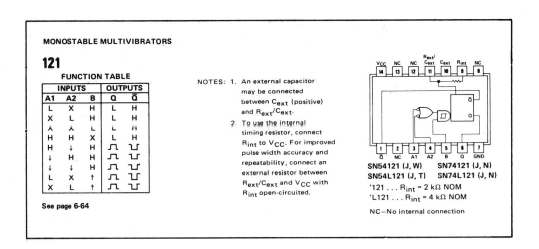

Figure 8.3(a) Monostable truth table and pin diagram for 74121.

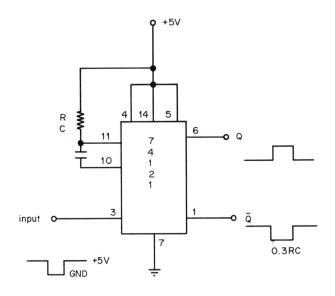

Figure 8.3(b) 74121 Wiring Diagram.

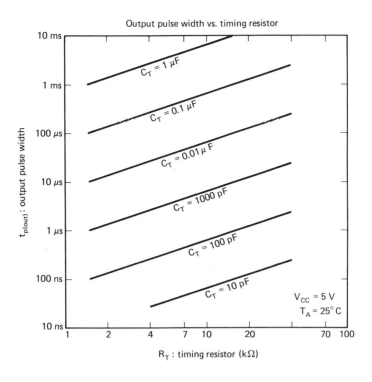

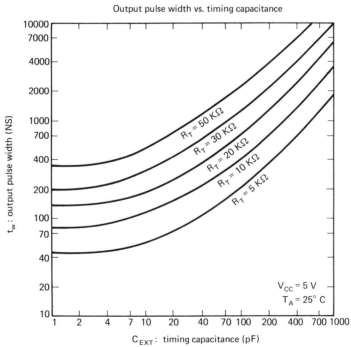

Figure 8.3(c)   One-shot timing graphs.

8-8   WAVE GENERATION AND WAVESHAPING

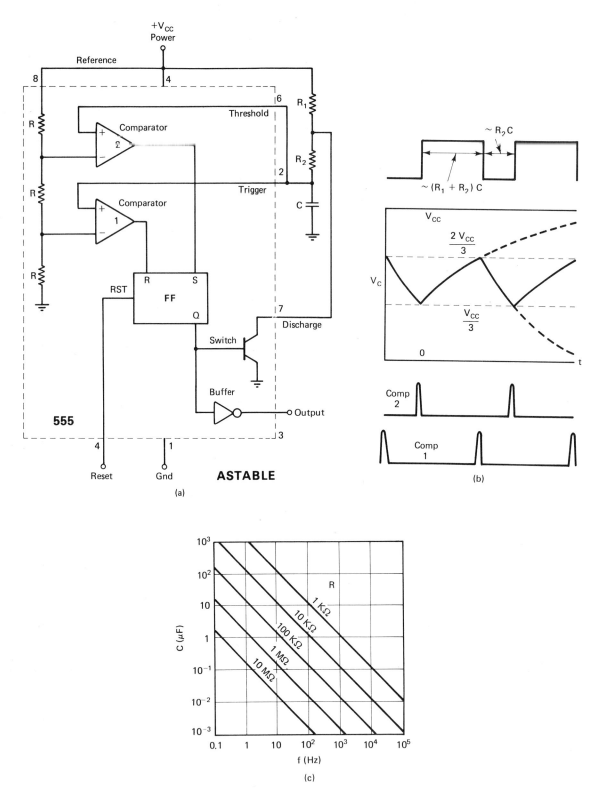

Figure 8.4 Timer as astable multivibrator.

WAVE GENERATION AND WAVESHAPING 8-9

4. Integrated Circuit Astable Multivibrator

Hook up a 555 timer as an astable multivibrator, using the circuit shown in Fig. 8.4.

Vary R and C each over several decades to verify <u>qualitatively</u> how the frequency varies. A frequency counter is desirable, although an oscilloscope can be used to measure the frequency if necessary. Now with C constant, vary R linearly over one decade. Plot the output frequency as function of R and compare with what you expect.

Once "calibrated," the 555 astable circuit is useful as a digital capacitance meter, using a frequency counter as an output. Take 10 apparently identical capacitors. Measure the frequency when each one is used in turn. What kind of a distribution of frequencies is observed? Compare this with the precision expected for the capacitors used.

(Optional but fun) Select one of the 555 applications shown in Text Section 6.4, or in the Signetics application notes or in Lancaster's <u>TTL Cookbook</u>. Try it out, and describe your observations.

EXPERIMENT 9: A COUNTER MEASUREMENT OF ELAPSED TIME

A. BACKGROUND

This experiment takes a jump into a digital measurement example, which includes many of the devices explained in detail in the previous two sections experiments. Now we show you how to use them in an interesting application, which also shows how simple it is to make a digital measuring instrument of remarkable precision. The diagrams in text Fig. 6.16, show how to measure the time interval between two interruptions of a light beam. The events we will use in the lab will be down-to-earth such as running speed or falling bodies. In scientific applications, the same ideas and similar circuits measure much shorter time intervals such as ballistics or cosmic rays.

B. REFERENCES

Text, Chapter 6, Section 4.

C. PROBLEM

The circuit in Fig. 9.1 is intended for measuring the time it takes an object to go from point **A** to point **B**. Suppose the object is a falling body. Can you think of a simple circuit modification which would allow a measurement of <u>instantaneous</u> velocity at the single point **B**? Hint: Use an object of well-defined and measureable size.

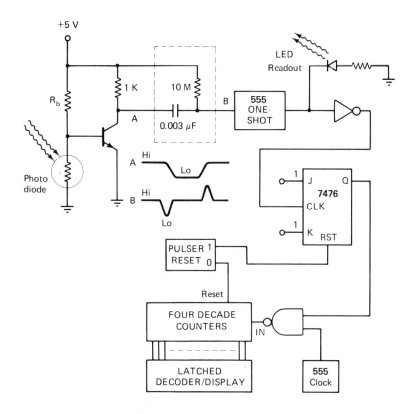

Figure 9.1 Circuit diagram for digital period measurement.

D. EQUIPMENT

Two **breadboard** sockets (one for counter, one for waveshaping)

5 V Power supply. DO NOT USE BATTERY POWER FOR THE COUNTER. The 4-digit LED display draws about 0.5A!

Four **decimal displays** (either 7-segment plus decoder on outboard, or latched with built-in decoder driver)*

Four **7490 decimal counter IC's**

---

* If you have a unit such as the Hewlett-Packard 5082-7340 with transparent plastic encapsulation, you can get a view of the operating LED matrix, and of the IC decoder-driver, using a low (10-50X) power microscope.

9-2   MEASUREMENT OF ELAPSED TIME

One **7400 NAND** or **7408 and IC**

One one-shot (**monostable multivibrator**) IC. Either a 555 timer or a 74121 monostable will do.

One **flip flop** IC (Example: 7474)

One **photo-detector** (phototransistor, or photodiode plus transistor)

One Laser (low power) or other collimated light source
Oscilloscope

E. MEASUREMENTS I: Event Timing and Waveshaping

Set up the waveshaping electronics shown in Fig. 9.1. Test all blocks separately before interconnecting. All waveshaping electronics will fit on a single breadboard socket. The counter and display will go on a second breadboard socket (later). The detector may take some adjustment of components or even of the circuit before an adequate (> 2 V) transition is obtained.

For the monostable, a 555 timer, (Fig. 6.10) will be adequate. The ON-time is not critical: long enough to trigger the flip flop (that's guaranteed) and not so long as to possibly overlap with the second "event." 10-100 msec seems reasonable. One could use instead a monostable such as the 74121. With the 74121, one chooses between two modes: an edge-triggered mode (pins 3 or 4, with pin 5 high; triggers on a negative edge) or a level-sensitive mode or a level-sensitive mode (pins 5 going high, with pins 3 and 4 held high). The second mode might be more reliable if the photodetector output is ragged (beware of multiple firings <u>within</u> a single "event," and check with a scope if needed).

For the flip flop, a variety of circuits is possible. One wants a device which changes its state each time a pulse is received. One way is shown in Fig. 9.1, using a JK flip flop in a toggle mode. The Set and Clear inputs are useful in establishing the correct state prior to a timing measurement.

Once all modules are working, connect them together and test the complete instrument. Note the waveforms you observe on a scope when the light beam is interrupted.

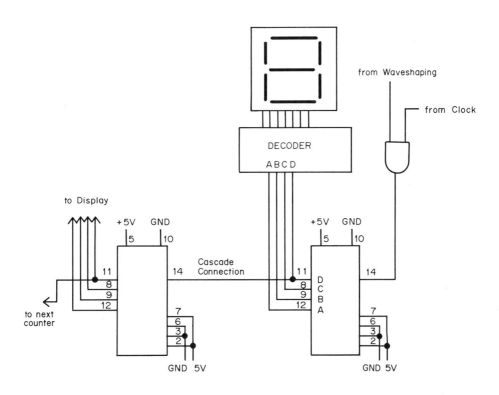

Figure 9.2 Connections for cascading 7490 counters.

MEASUREMENTS II: Counter Section

1. Display

Latched Display (Option I). Explore the outputs of the display using a 4-bit switch register, and 0 or 5 V logic levels controlling the latch and decimal point functions. Verify that latched information remains displayed even when the input is later changed.

9-4   MEASUREMENT OF ELAPSED TIME

Decimal Display (Option II). A latched decoded display is expensive and in fact unnecessary for this experiment. A seven-segment decimal display, with decoder driver (as on the outboards) is quite adequate. You will need 4-side-by-side to read the contents of the 4-digit counter. Turn off the display when not needed, since 4 digits draw nearly 0.5A!

2. Counter. Wire up a 4-digit counter using 4 7490's, as shown in Fig. 9.2. Do this digit by digit, verifying that each stage is working before wiring in the next. Use binary readout lights to check the outputs, and a slow clock (~1 HZ) to step the counter.

Driving sequential stages. The 8's output of stage n drives the input of state (n+1). This works only because of two properties of the 7490 (see Text, Fig. 5.7(a)).

a. The count sequence is BCD decimal rather than full 4-bit binary counter. What does the 8's or "D" output do between decimal 9 and decimal 0?

b. The counter changes states following a negative going transition (negative edge).

How do (a) and (b) facilitate a multiple digit decimal counter?

With all 4 stages of input connected [(8's bit out)$_n$ to (input)$_{n+1}$], verify with binary readout lights that all stages correctly divide by 10, and that the correct 0-9 sequence is observed within each stage. Now wire the 1248 outputs of each stage to the 1248 inputs of the Numeric displays. [Option I only: Wire the latch inputs in parallel to a switch to create a latched (5 V) or active (0 V) mode.] Verify full 4-digit operation using a variable speed clock; 100 kHz on the input should increment the most significant digit how often?

3. Counter Measurements. Wire up an AND gate to allow the counter to be started and stopped (Fig. 9.2). You now have the capability to time events precisely. Two things are necessary first:

   a. The start-stop operation must be reliable and repeatable.

   b. The clock must be calibrated.

Start/Stop gating and switch bounce. It matters what is used to enable the counter. An ordinary contact closure in fact bounces between open and closed on the way to a stable state. Since TTL logic is fast enough to see those unwanted events, an ordinary switch is not suitable to enable the counter.

Explore contact bounce with the following inputs connected to pin 14 of the least significant digit:

   i. A wire being inserted into a ground terminal on the breadboard.

   ii. An ordinary slide switch being connected to ground.

   iii. A "debounced" pulser.

Take enough readings of the increment in counts observed to build up a picture of the reliability of each.

Counter Calibration. For our purposes, approximate calibration is adequate. The stability of the clocks we will use is not outstanding, and the emphasis will be on relative times. In precision applications, crystal clocks are used, with 6-digit stability.

Clock speed: For the measurement of "ordinary" **events** (falling bodies, running speed) a useful 4-digit counter is

   10's   1's   .1's   .01's
   sec   sec   sec   sec

with a full range of 99.99 sec.

Clock selection: If you are using a fixed clock such as on a logic trainer, continuous adjustment of clock rate may not be possible. Select the rate which results in the closest approximation to the above. Calibrate the readout (10,000 counts = N sec) using a stopwatch. If you are using a variable clock, such as a 555 timer, the counter readout may be adjusted using appropriate parallel capacitances. In either case, use an AND gate to set the ON period.

MEASUREMENTS III: Interfacing Counter to Detector

1. Connect the counter to the photodetector-waveshaping circuit, and measure the speed of various events (running, bicycle, baseball, falling body . . .). Use the 7490 reset (pin 2 or 3) to initialize the counter to zero. Note how the high resolution made possible by <u>digital</u> measurement allows a precise comparison of events.

2. (Optional) <u>Reaction Time Experiment</u>. Have one person start counter (unknown to 2nd person). The 2nd person tries to turn it off as soon as a count is observed. Average the results to determine the reaction time to a precision of $\simeq 0.1$ sec.

3. This part of the experiment is quite open-ended. There are many physical systems you might explore. A few interesting possibilities will be mentioned.

    a. If you pass your hand in the beam so the fingers interrupt the light, does it generate one pulse or more? Multiple pulses could be understandable if one wants to measure the hand's travel time from points A to B. Can you adjust the 555 on-time to take care of this? Note that multiple pulses <u>could</u> be used as a truly "digital" (fingers) instantaneous velocity measurement. How?

    b. Look at and photograph the waveforms on a scope at the following points in the circuit:
    Photodiode output
    Transistor output
    One shot output
    Flip flop output
    NAND output
Comment on the digital waveshaping that goes on.

BE ON THE ALERT FOR PROBLEMS WITH THE 555 TIMER.
    If the counter continues counting when it should be stopped, the chances are that it is due to the "retriggering" of the 555. If the input goes high during the output pulse, the output will not fall to zero at the end of the expected time.

c. Measure the period of a pendulum, and compare with what you expect [$2\pi f = (g/l)^{1/2}$]. Now measure the instantaneous velocity at the bottom of the swing. How?

d. Measure the travel time for a disk to roll down a slope. Use two disks of the same mass but different moment of inertia, and compare the times.

EXPERIMENT 10: SHIFT REGISTERS

A. BACKGROUND

Shift registers are used to conveniently store and manipulate data. One application is in parallel to serial data communications, since it's easier to connect one wire between two locations than n wires. Any digital transmission via phone lines works this way, and shift registers are what do the job. Another application is in computation, since a shift right of an arbitrary binary number is equivalent to multiplying it by a power of two.

We will explore one example of a shift register, the 74194. The 74194 is very flexible: it can parallel load, shift in either direction, and is synchronous. The fact that it has four bits makes binary readout simple to implement with 4 lights or a seven-segment display. However, 4 bits makes a limited vocabulary, and the student seriously interested for example in ASCll communication may wish to try using instead an 8-bit unit (74164 and 74165).

B. REFERENCES

For simpler cookbook experiments on this subject, see Larsen & Rony, Logic and Memory Experiments using TTL Integrated Circuits.

C. PROBLEMS

1. Text, Problem 5.8.
2. Sketch the internal circuit of the 74194 shift register, text Fig. 5.18(e); see also 74194 specifications below (Fig. 10.1). Explain the function of each block. In particular, explain how the truth table for the mode control is implemented.

TTL
MSI

TYPES SN54194, SN54LS194A, SN54S194, SN74194, SN74LS194A, SN74S194
4-BIT BIDIRECTIONAL UNIVERSAL SHIFT REGISTERS
BULLETIN NO. DL-S 7611866, MARCH 1974–REVISED OCTOBER 1976

- Parallel Inputs and Outputs
- Four Operating Modes:
  - Synchronous Parallel Load
  - Right Shift
  - Left Shift
  - Do Nothing
- Positive Edge-Triggered Clocking
- Direct Overriding Clear

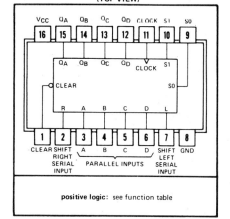

SN54194, SN54LS194A, SN54S194 . . . J OR W PACKAGE
SN74194, SN74LS194A, SN74S194 . . . J OR N PACKAGE
(TOP VIEW)

positive logic: see function table

TYPE	TYPICAL MAXIMUM CLOCK FREQUENCY	TYPICAL POWER DISSIPATION
'194	36 MHz	195 mW
'LS194A	36 MHz	75 mW
'S194	105 MHz	425 mW

description

These bidirectional shift registers are designed to incorporate virtually all of the features a system designer may want in a shift register. The circuit contains 46 equivalent gates and features parallel inputs, parallel outputs, right-shift and left-shift serial inputs, operating-mode-control inputs, and a direct overriding clear line. The register has four distinct modes of operation, namely:

Parallel (broadside) load
Shift right (in the direction Q_A toward Q_D)
Shift left (in the direction Q_D toward Q_A)
Inhibit clock (do nothing)

Synchronous parallel loading is accomplished by applying the four bits of data and taking both mode control inputs, S0 and S1, high. The data are loaded into the associated flip-flops and appear at the outputs after the positive transition of the clock input. During loading, serial data flow is inhibited.

Shift right is accomplished synchronously with the rising edge of the clock pulse when S0 is high and S1 is low. Serial data for this mode is entered at the shift-right data input. When S0 is low and S1 is high, data shifts left synchronously and new data is entered at the shift-left serial input.

Clocking of the flip-flop is inhibited when both mode control inputs are low. The mode controls of the SN54194/SN74194 should be changed only while the clock input is high.

FUNCTION TABLE

INPUTS										OUTPUTS			
CLEAR	MODE		CLOCK	SERIAL		PARALLEL				Q_A	Q_B	Q_C	Q_D
	S1	S0		LEFT	RIGHT	A	B	C	D				
L	X	X	X	X	X	X	X	X	X	L	L	L	L
H	X	X	L	X	X	X	X	X	X	Q_{A0}	Q_{B0}	Q_{C0}	Q_{D0}
H	H	H	↑	X	X	a	b	c	d	a	b	c	d
H	L	H	↑	X	H	X	X	X	X	H	Q_{An}	Q_{Bn}	Q_{Cn}
H	L	H	↑	X	L	X	X	X	X	L	Q_{An}	Q_{Bn}	Q_{Cn}
H	H	L	↑	H	X	X	X	X	X	Q_{Bn}	Q_{Cn}	Q_{Dn}	H
H	H	L	↑	L	X	X	X	X	X	Q_{Bn}	Q_{Cn}	Q_{Dn}	L
H	L	L	X	X	X	X	X	X	X	Q_{A0}	Q_{B0}	Q_{C0}	Q_{D0}

H = high level (steady state)
L = low level (steady state)
X = irrelevant (any input, including transitions)
↑ = transition from low to high level
a, b, c, d = the level of steady-state input at inputs A, B, C, or D, respectively.
$Q_{A0}, Q_{B0}, Q_{C0}, Q_{D0}$ = the level of Q_A, Q_B, Q_C, or Q_D, respectively, before the indicated steady-state input conditions were established.
$Q_{An}, Q_{Bn}, Q_{Cn}, Q_{Dn}$ = the level of Q_A, Q_B, Q_C, respectively, before the most-recent ↑ transition of the clock.

TEXAS INSTRUMENTS
INCORPORATED

Figure 10.1 74194 Specifications sheet.

EQUIPMENT
    Breadboard Socket
    Outboards: pulser
             switch register
             seven-segment display
             4-bit binary lights
    74194   4-bit shift register

    Optional: (in addition)
             voltage controlled oscillator
             555 timer clock
             scope
             imagination

    Alternative: 74164 and 74165 8-bit shift register. You will need one of each to complete the lab, and a double set of switch registers and readout lights. The pin connections will <u>not</u> be as shown in Fig. 10.2 and 10.3. Consult Larson & Rony if you do the 8-bit version.

E. MEASUREMENTS
    <u>**1. Basic Shift Register.**</u>
    a. Wire up the circuit shown in Fig. 10.2. Note that the two <u>mode</u> controls the SR function. Their truth table allows them to inhibit, to shift right, to shift left, and to parallel load. Connect them to 1 or 0 depending on the desired function. The clear input is normally tied high. Momentarily bring it to 0 to clear. The serial inputs (Left and Right) are brought sequentially high or low to enter words desired when in the <u>Serial</u> mode.
    Put binary readout lights in parallel with the seven-segment display to aid in decoding. Note that the pulser is wired normally <u>high</u>.

    b. Explore the parallel load operation ($S_0 = S_1 = 1$). Set various numbers in the switch register, and clock them into the parallel inputs. Verify that any 4-bit word can be entered. Record your binary inputs and observed binary outputs. (Left and Right serial input connections irrelevent here).

SHIFT REGISTERS   10-3

c. Explore serial loading from right ($S_0 = S_1 = 1$). Enter a string of 1's (pin 7 high). Observe and record the sequence of outputs at each clock pulse until no further change is observed. <u>Interpret</u> your results in terms of the binary contents of the register at each clock pulse. Why is no further change observed upon clocking?

d. With the contents still all 1's, wire the input low. Observe and record the sequence of outputs at each clock pulse until no further change is observed. Interpret the numbers you observed.

e. With the contents now all 0's, put the input high for one clock pulse. Then put it low again. Cycle the clock. Observe, record, and interpret the numbers you observe.

f. Change the mode to shift right, and verify that you can enter numbers from the least significant end. Repeat steps like c, d, e, above.

2. <u>Circulating Ring Registers.</u> How could you keep the numbers from disappearing out the ends on successive serial operations? Wire the least significant output to the left serial input, and wire the most significant output to the right serial input. Parallel load a number via the switch register. Change the mode to shift right or left. Verify that your number can be made to circulate around the register in either direction.

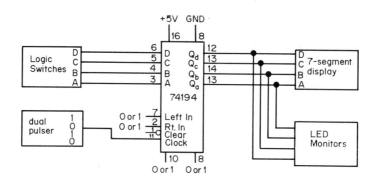

Figure. 10.2 Shift register test circuit.

3. Parallel to Serial to Parallel data transmission.

a. This will take cooperation with another student since basically two complete input/output arrays are needed. Wire up the circuit shown in Fig. 10.3. Note that parallel entry is possible from either "port," and data transmission may proceed in either direction depending on the mode selection. One can either imagine this as a way to cascade n-bit registers, or as a way to transmit parallel information in serial form. (You may wish to have the two "ports" in two locations such that one observer can only see his/her own parallel readout lights.)

b. Verify that 8-bit parallel loads are possible, and that the clear operation works.

c. Set the mode to transmit to the left, and parallel load a word into the right port. Cycle the clock four times and verify that the word has been received at the right port.

d. **Clear.** Set the mode to transmit to the left, and parallel load a word into the right port. Cycle the clock four times and verify that the word was received.

e. (Optional: only if you are using 74164 and 74165.) You now, of course, can send and receive meaningful ASCII messages. Using the ASCII code, send a message to your friend across the room.

4. (Optional) Serial digital audio communications. Wire up the shift register as a circulating ring. Use as a clock a 555 timer set at some convenient (0.1-1 Hz) rate. Connect any single output bit to the input of a voltage controlled oscillator. (A voltage divider may be needed to bring the due to the 5v TTL transition to a reasonable value.) Observe on a scope or if possible with an audio output the "frequency shift keying," as the telephone company calls it. Verify that you can parallel load and then transmit any 4-bit word as a string of audio tones.

5. (Even more optional) Digital RF communications. The output of that voltage controlled oscillator could be used to modulate an RF carrier, and thus broadcast the serial information to any radio set to the carrier frequency. You could do it with a multiplier or another modulator, and any RF oscillator, plus a wire to act as an antenna for short distance communication (within a room).

6. (Far out optional). If you are using the 8-bit units, consider the possibility of talking to or from a teletype. Don't undertake this unless you already know something about ASCII and various communications standards (EIA, etc.). You can easily build a 110 baud (bits per sec.) clock using a 555 timer.

CAUTIONARY NOTE:
Here are some suggestions to avoid garbage getting into your data.
   a. When changing the MODE connections, or when parallel loading, the clock must be kept high, or garbage results.

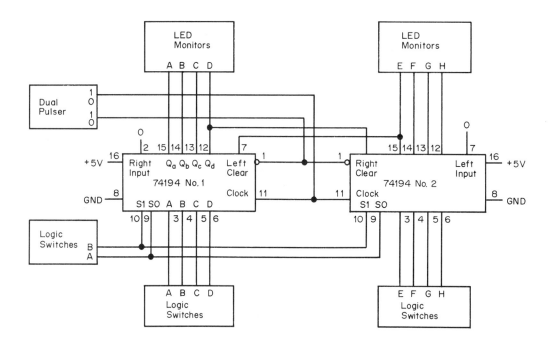

Figure. 10.3 Parallel to serial and serial to parallel synchronous communication circuit.

b. Shift registers are sensitive to having the correct power supply voltage. For example, a ring counter will give garbage out if the power supply voltage is low. In the serial communications part of the lab, using two shift registers, both power supplies must be 5v or garbage will come out.

EXPERIMENT 11: MEMORY

A. BACKGROUND

This experiment makes use of a small (16 4-bit words) random access memory (RAM) chip (the 7489) to demonstrate basic RAM features. Some are basic to any RAM (the distinction between address and contents) and some are peculiar to the types of semiconductor memory called Volatile, which lose the data when the power is turned off. Even though the 7489 is obsolete, except for occasional test signal processing applications, it fits in so well with the 4-bit structure of the outboards that we retain its use to illustrate memory concepts.

A 4 to 10 decoder is then driven by the memory, demonstrating that the contents can cause a specific line to be selected or not selected during the 16 cycles of the counter.

That same decoder is then used to reset the counter, producing a simple example of a "program loop" which sequences through an arbitrary but selectable number of "program steps." The contents drives the decoder, and either resets or does not depending on whether a correspondence exists between the contents and the decoder output used to reset the counter.

B. REFERENCES
Text, Chapter 7.
Intel, Memory Design Handbook.
Texas Instruments, TTL Databook.

C. PROBLEMS
Text, 7.2, 7.3, and 7.4.

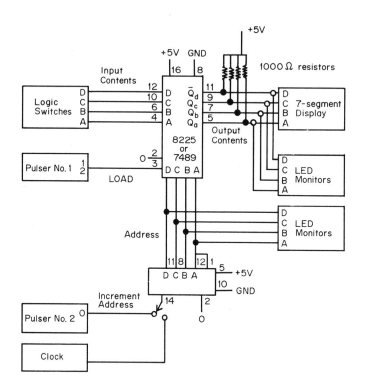

Figure 11.1 Memory test circuit.

D. **EQUIPMENT**
   Breadboard
   Pulser
   Clock
   7-segment display
   4-bit LED monitor, (a second one is desirable also)
   7489 4 x 16 RAM [Refer to Text, Table 7.1 for specifications and circuit]
   7493 4-bit binary counter
   7442 4 to 10 decoder
   7404 Inverter

E. MEASUREMENTS

1. Memory.

a. Wire up the circuit shown in Fig. 11.1. If possible, put a second 4-bit lamp monitor in parallel with the 7-segment display, to aid in interpreting the results (the memory output of this chip is the complement of the input).

b. Cycle the counter through all 16 states. Record the initial contents of each memory location.

c. Reset the address to location zero (pin 2 of the 7493 momentarily high). Explore what happens when you:

Load 1111   Output =   ?   Why?
Load 0000   Output =   ?
Load   ?   Output = 0011

NOTE: "Load" refers to the binary switch register contents. The output should be its complement with this particular memory chip. The word **Store** will be used to refer to loading a desired **output**.

d. Fill the entire memory with a number sequence. For exmple, use your telephone number, with area code and blanks. Pick the input numbers to produce a desired output sequence. Now cycle the counter through all 16 states to verify that what you put in got there, and is still there at the correct address.

e. Turn the power off and then on again. What is the contents of all 16 locations now? This kind of memory is called volatile, and you have just seen why.

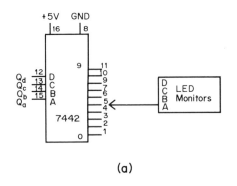

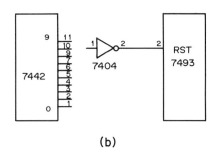

Figure 11.2 (a) Decoding a memory data output; (b) Using decoding to reset a memory address.

2. Using memory contents to drive a decoder.

a. Wire the circuit shown in Fig. 11.2(a), without disturbing the existing memory circuit. First verify the decoder operation by using as inputs simple 0 V and 5 V voltage levels. Recall: the **selected** output is **low**; all others are high.

b. Now let the decoder input be the memory output. Store 0000 everywhere in memory. Which decoder output is selected? In how many of the 16 cycles?

c. Store a 0011 in memory location 0101, and 0000 elsewhere. Cycle the memory and observe which 7442 output is selected and <u>when</u>.

The **Contents** determines <u>Which</u>
The **Address** determines <u>When</u>

d. Store in the memory an arbitrary number sequence such as 1, 7, 3, 8, 12, 2, 15, 5... Cycle the memory and observe which output of the 7442 is selected and when. You have now made a non-sequential decoder, which selects any desired sequence of lines depending on the **Contents** of memory. Suppose the number 4 is not among those in memory. Will line 4 be selected during the 16-cycle sequence?

3. Using memory contents plus a decoder to make a "program loop."

a. Add to your previous circuit an inverter, tying a selected 7442 output to the 7493 reset [Figure 11.2(b)]. The result of having that line selected by the memory **contents** will be to immediately reset the counter (hence the memory **address**) back to zero.

b. With the 7493 Reset <u>**disconnected**</u> (wire to 0), store a 16-number sequence of small numbers (1,2,3,4, for example). Wire the 7493 Reset to a high number output (7, 8, 9) of the 7442. Do you expect to be able to sequence through all 16 memory locations? Try it.

c. Replace one of those numbers in memory with a large number such as 7 at a memory address such as 12. (Wire the 7493 Reset to 0 while doing this.) Now connect the 7442's output 7 to the 7493 Reset (after an inversion). Sequence through the memory addresses. What <u>range</u> is now accessible? Which determines it: the address or the contents?

By the way, the Reset command requires that 7 be a memory output, but did you see that 7 appear on the readout lights? Explain. In fact, did you see address 12 appear? Sequence it again and explain the reset cycle.

MEMORY   11-5

d. Explore various changes. What happens if you wire the Reset to a different decoder address? What happens if you change the contents at the address previously used as a reset? You have now made a simple "program loop," where the contents of a given address causes (or doesn't cause) execution of a "return to starting address."

TTL LSI

TYPE SN7489
64-BIT READ/WRITE MEMORY

- For Application as a "Scratch Pad" Memory with Nondestructive Read-Out
- Fully Decoded Memory Organized as 16 Words of Four Bits Each
- Fast Access Time . . . 33 ns Typical
- Diode-Clamped, Buffered Inputs
- Open-Collector Outputs Provide Wire-AND Capability
- Typical Power Dissipation . . . 375 mW
- Compatible with Most TTL and DTL Circuits

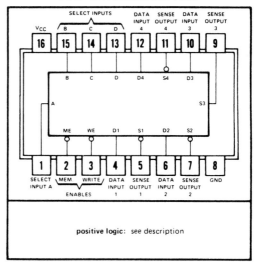

J OR N DUAL-IN-LINE OR W FLAT PACKAGE (TOP VIEW)†

positive logic: see description

†Pin assignments for these circuits are the same for all packages.

description

This 64-bit active-element memory is a monolithic, high-speed, transistor-transistor logic (TTL) array of 64 flip-flop memory cells organized in a matrix to provide 16 words of four bits each. Each of the 16 words is addressed in straight binary with full on-chip decoding.

The buffered memory inputs consist of four address lines, four data inputs, a write enable, and a memory enable for controlling the entry and access of data. The memory has open-collector outputs which may be wire-AND connected to permit expansion up to 4704 words of N-bit length without additional output buffering. The open-collector outputs may be utilized to drive external loads directly; however, dynamic reponse of an output can, in most cases, be improved by using an external pull-up resistor in conjunction with a partially loaded output. Access time is typically 33 nanoseconds; power dissipation is typically 375 milliwatts.

FUNCTION TABLE

ME	WE	OPERATION	CONDITION OF OUTPUTS
L	L	Write	Complement of Data Inputs
L	H	Read	Complement of Selected Word
H	L	Inhibit Storage	Complement of Data Inputs
H	H	Do Nothing	High

write operation

Information present at the data inputs is written into the memory by addressing the desired word and holding both the memory enable and write enable low. Since the internal output of the data input gate is common to the input of the sense amplifier, the sense output will assume the opposite state of the information at the data inputs when the write enable is low.

read operation

The complement of the information which has been written into the memory is nondestructively read out at the four sense outputs. This is accomplished by holding the memory enable low, the write enable high, and selecting the desired address.

Texas Instruments
INCORPORATED

Figure 11.3 7489 specifications sheet.

EXPERIMENT 12:   BINARY ADDERS

A.   BACKGROUND

In this experiment, you will construct a 1-bit full adder to get the idea of what the truth table looks like and why.

Then we leap directly to a full 4-bit adder. It takes 8 exclusive OR's and 12 NANDS, and a lot of wiring. The same thing is available as one 16-pin package (7483), but it's perhaps worth doing it once with gates to convince yourself that logic really works.

Those of you who would rather accept the previous statement, and move on to something more intriguing, should ignore all of the instructions in Version I and instead explore the truth table of the 74181 Arithmetic and Logic Unit (Version II).

B.   REFERENCES

Text, Chapter 8.

Texas Instruments: Designing with TTL Integrated Circuits, Ch. 9: Arithmetic Elements.

For the **subtraction** part, look up your favorite computer science text on the representation of negative numbers.

C.   PROBLEMS

Text, Problems 8.1, 8.2, and 8.3.

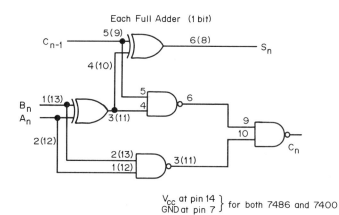

Figure 12.1 Logic and wiring diagram for the full adder from TTL IC's. The numbers in parenthesis are alternate gates on the same package.

D. EQUIPMENT
   1 **Breadboard**
   IC's
        (for option I):
            2 7486 quad exclusive OR
            3 7400 quad NAND
        (for option II):
            74181 ALU
   Outboards
        2 4-bit switch registers
        1 4-bit binary readout light
   Lots of wires
   Lots of patience
The adder will all fit on one socket if the two switch registers are opposite one another.

E. MEASUREMENTS

1. Wire up a 1-bit full adder as shown in Fig. 12.1. The "preceding" carry bit C_{n-1} will be a wire to 0 or 1. Use two switches to encode A and B inputs, and 2 binary lights to read the sum **S** and the carry **C**.

Verify the truth table (Text, Table 8.2). Now select **Version I** or **Version II**.

Version I.

2. Wire up a 4-bit full adder as shown in detail in Fig. 12.2. Be very careful and patient; work out a scheme to keep track of which wires you have put in place. Use two 4-bit switch registers for **A** and **B**, and a 4-bit binary light to read **S**. C_0 will be a wire to **0** or **1**. C_4 will go nowhere unless you have a fifth light handy.

The chances are that it will **not** work **the first time**. To debug it, take out the carry wires between stages, replace with a wired carry, and test each 1-bit full adder separately, using a light to read S and one to read C. When all four stages work, replace the carry interconnections.

Verify that it will indeed correctly add any two 4-bit numbers.

3. (**Optional**, but recommended for computer science majors and the like). Explore binary subtraction. Use 2's complement notation, i.e., for an N-bit number,

$$(-X) + (X) = 2^N$$
$$(-X) = 2^N - X$$

Example: $N = 4$, $2^4 = 16_{10} = 10000$
$$2^N - 1 = 1111$$

With $\quad 1_{10} = 0001,$
Then $\quad -1_{10} = 1111$

Note that $1 + (-1) = 0$
```
   0001
 + 1111
   0000
```

The carry is suppressed. Note that the MSB becomes the sign bit; 1xxx is a negative number.

BINARY ADDERS 12-3

RULE
To make a 2's complement number, make the exact complement and add a 1.

The negative of 1010 is (0101 + 0001) = 0110.

Version II.
Explore the truth table of the 74181 ALU, (Text, Table 8.4). The functions are selected by the 4-bit Select Inputs. These are two kinds of functions, depending on the state of the **Mode Control** input **M**. With **M = Low**, the 74181 performs one of 16 **arithemetic operations,** such as Add, Subtract, Shift, and Relative Magnitude. With **M = High**, the 74181 functions as a programmable **Logic chip,** performing such functions as AND, OR NAND, NOR, XOR, and COMPARE.

4. **(Optional)** For either Version I or Version II. (Note: Don't bother with this if you also plan to do Exp. 13). Did you save your wired-up memory board from Experiment 11? If so, explore the following challenge:

Design a circuit which will take the contents of memory location 1, add it to the contents of memory location 2, and put the result in memory location 3. Try it. Congratulations! You have made a computer. (Obtain extra IC's as needed from the TA. For sure, you will need the equivalent of two 4-bit registers (Latches).)

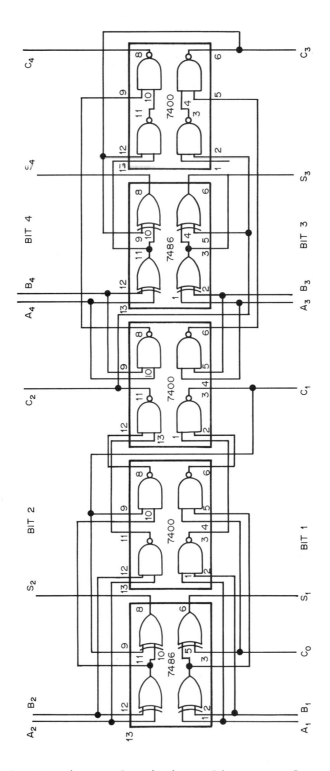

Figure. 12.2 Logic and wiring diagram for a 4-bit full-adder from TTL MSI IC's. This is only for the brave, stubborn, or foolhardy; the same logic function is all in a single 7483 chip.

BINARY ADDERS   12-5

EXPERIMENT 13:

PUTTING IT TOGETHER; A TTL MSI 4-BIT MICROCOMPUTER

A. BACKGROUND

This is a project lab, intended only for computer scientists, or others interested in designing computer architecture. No one in their right mind would go through the work below to build the equivalent of a microprocessor which can be bought for $10. The experiment is quite open-ended in its goals, since the details of architecture are left as part of the project. The purpose is to show how the memory, arithmetic, and register elements available in MSI TTL can be combined to process a 4-bit slice or "nibble" of information. Such 4-bit arrays can readily be cascaded to arbitrary word length. This **bit slice method** is one of the ways in which modern computer design utilizes improved logic elements to emulate an older machine without changing the external architecture or programming seen by the user.

Although these particular TTL elements are being superseded by newer (particularly LSI) building blocks, TTL logic is still among the fastest of logic elements, and still useful in prototype breadboarding.

Two different arithmetic units are available. Select **one**. The 7483 is a four-bit full adder, with internal carry. The 74181 is called an **arithmetic and logic unit (ALU)**, including addition, subtraction, shifting, and many Boolean logic functions. Specifications for these two IC's are given in Fig. 1 and 2. (Warning: the 7483 costs only $2, but the 74181 costs $10.)

Design a basic computer to operate on two four-bit numbers. Minimal requirements are two registers (A and B) to store the inputs, and some place to put the result. (What kinds of FF's should be used as latches?) Memory is

useful for storing both inputs and results. Use a 7489.
How will you route an input to the correct register? How
will you get a result into memory? How will the "program"
sequence get the right thing from the right place at the
right time (cycle)?

Other possible design considerations: If you are using
a 7481 and want to make a "Boolean calculator," how would
you make it so the operator could select the function
performed with a single keystroke rather than a binary word?
(The solution involves 16 to 4 decoding.) What about the
display? Binary LED's would be best, but how many sets; one
for each register, or should they be multiplexed? How
should numbers be entered? Binary is simplest, but again
one might prefer a keystroke approach with some decoding.

With such an open-ended project, **avoid** getting in so
deep that when you build it and it doesn't work, debugging
is a great task. Do the complete design, specifying the
goals clearly and how you plan to implement them. When
building it, however, do not go directly to the final
design. Test each section separately. For example, it
would be worthwhile testing the arithmetic unit with simple
switch register inputs before adding memory.

It would be advisable to have a discussion group to go
over designs and find bugs before building them in. Also,
check about the availability (and cost!) of parts for your
design.

B. REFERENCES

Text, Chapter 8.

Texas Instruments, TTL Databook, especially on the 74181.

TTL MSI
TYPES SN5483A, SN54LS83, SN7483A, SN74LS83
4-BIT BINARY FULL ADDERS

- For applications in:
  Digital Computer Systems
  Data-Handling Systems
  Control Systems

- SN54283/SN74283 Are Recommended For New Designs as They Feature Supply Voltage and Ground on Corner Pins to Simplify Board Layout

TYPE	TYPICAL ADD TIMES TWO 8-BIT WORDS	TWO 16-BIT WORDS	TYPICAL POWER DISSIPATION PER 4-BIT ADDER
'83A	23 ns	43 ns	310 mW
'LS83	89 ns	165 ns	75 mW

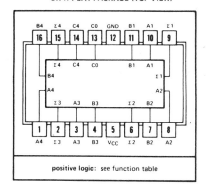

positive logic: see function table

FUNCTION TABLE

INPUT				OUTPUT WHEN C0 = L			WHEN C0 = H		
				WHEN C2 = L			WHEN C2 = H		
A1 / A3	B1 / B3	A2 / A4	B2 / B4	Σ1 / Σ3	Σ2 / Σ4	C2 / C4	Σ1 / Σ3	Σ2 / Σ4	C2 / C4
L	L	L	L	L	L	L	H	L	L
H	L	L	L	H	L	L	L	H	L
L	H	L	L	H	L	L	L	H	L
H	H	L	L	L	H	L	H	H	L
L	L	H	L	L	H	L	H	H	L
H	L	H	L	H	H	L	L	L	H
L	H	H	L	H	H	L	L	L	H
H	H	H	L	L	L	H	H	L	H
L	L	L	H	H	L	L	L	H	L
H	L	L	H	L	H	L	H	H	L
L	H	L	H	L	H	L	H	H	L
H	H	L	H	H	H	L	L	L	H
L	L	H	H	H	H	L	L	L	H
H	L	H	H	L	L	H	H	L	H
L	H	H	H	L	L	H	H	L	H
H	H	H	H	H	L	H	L	H	H

H = high level, L = low level

NOTE: Input conditions at A3, A2, B2, and C0 are used to determine outputs Σ1 and Σ2 and the value of the internal carry C2. The values at C2, A3, B3, A4, and B4 are then used to determine outputs Σ3, Σ4, and C4.

description

These full adders perform the addition of two 4-bit binary numbers. The sum (Σ) outputs are provided for each bit and the resultant carry (C4) is obtained from the fourth bit. The adders are designed so that logic levels of the input and output, including the carry, are in their true form. Thus the end-around carry is accomplished without the need for level inversion. Designed for medium-to-high-speed, the circuits utilize high-speed, high-fan-out transistor-transistor logic (TTL) but are compatible with both DTL and TTL families.

The '83A circuits feature full look ahead across four bits to generate the carry term in typically 10 nanoseconds to achieve partial look-ahead performance with the economy of ripple carry.

The 'LS83 can reduce power requirements to less than 20 mW/bit for power-sensitive applications. These circuits are implemented with single-inversion, high-speed, Darlington-connected serial-carry circuits within each bit.

Series 54 and 54LS circuits are characterized for operation over the full military temperature range of $-55°C$ to $125°C$; Series 74 and 74LS are characterized for $0°C$ to $70°C$ operation.

absolute maximum ratings over operating free-air temperature range (unless otherwise noted)

Supply voltage, V_{CC} (see Note 1) . . . . 7 V
Input voltage . . . . 5.5 V
Interemitter voltage (see Note 2) . . . . 5.5 V
Operating free-air temperature range: SN54', SN54LS' Circuits . . . . $-55°C$ to $125°C$
                                      SN74', SN74LS' Circuits . . . . $0°C$ to $70°C$
Storage temperature range . . . . $-65°C$ to $150°C$

NOTES: 1. Voltage values, except interemitter voltage, are with respect to network ground terminal.
2. This is the voltage between two emitters of a multiple-emitter transistor. For the '83A, this rating applies between the following pairs: A1 and B1, A2 and B2, A3 and B3, A4 and B4. For the 'LS83, this rating applies between the following pairs: A1 and B1, A1 and C0, B1 and C0, A3 and B3.

TEXAS INSTRUMENTS
INCORPORATED

Figure 13.1 7483 4-bit full adder specifications.

C. PROBLEMS

1. Design your own computer as outlined above.

2. Explain the logic of the 7483 4-bit adder. How does the circuit carry out the truth table? (See Fig. 13.1)

3. The circuit for the 74181 (Fig. 13.2) contains 75 gates, and is sufficiently complicated that it will not be explored. Instead, consider the truth table for the operation performed as a response to a given combination of the four **select** inputs. Explain which ones will be essential for the principal operations of a computer. A **minimal set** is **add, subtract,** and **shift**.

4. Present the design for a TTL MSI computer. Explain the goals. How does your design implement those goals? What would it cost for 4-bit words and 16-bit words of memory? What would it cost when extended to 16-bit words and 4192 words of memory?

D. EQUIPMENT
As needed.

E. MEASUREMENTS
Get it working and see if your design goals have been accomplished.

TTL MSI

TYPES SN54181, SN54LS181, SN54S181, SN74181, SN74LS181, SN74S181
ARITHMETIC LOGIC UNITS/FUNCTION GENERATORS

PIN DESIGNATIONS

DESIGNATION	PIN NOS.	FUNCTION
A3, A2, A1, A0	19, 21, 23, 2	WORD A INPUTS
B3, B2, B1, B0	18, 20, 22, 1	WORD B INPUTS
S3, S2, S1, S0	3, 4, 5, 6	FUNCTION-SELECT INPUTS
$\overline{C_n}$	7	INV. CARRY INPUT
M	8	MODE CONTROL INPUT
F3, F2, F1, F0	13, 11, 10, 9	FUNCTION OUTPUTS
A = B	14	COMPARATOR OUTPUT
P	15	CARRY PROPAGATE OUTPUT
$\overline{C_{n+4}}$	16	INV. CARRY OUTPUT
G	17	CARRY GENERATE OUTPUT
V_{CC}	24	SUPPLY VOLTAGE
GND	12	GROUND

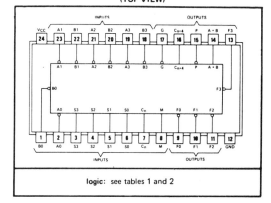

'181, 'LS181 . . . J, N, OR W PACKAGE
SN54S181 . . . J OR W PACKAGE
SN74S181 . . . J, N, OR W PACKAGE
(TOP VIEW)

logic: see tables 1 and 2

- Full Look-Ahead for High-Speed Operations on Long Words
- Input Clamping Diodes Minimize Transmission-Line Effects
- Darlington Outputs Reduce Turn-Off Time
- Arithmetic Operating Modes:
  Addition
  Subtraction
  Shift Operand A One Position
  Magnitude Comparison
  Plus Twelve Other Arithmetic Operations
- Logic Function Modes:
  Exclusive-OR
  Comparator
  AND, NAND, OR, NOR
  Plus Ten Other Logic Operations

TYPICAL ADDITION TIMES

NUMBER OF BITS	ADDITION TIMES			PACKAGE COUNT		CARRY METHOD BETWEEN ALU's
	USING '181 AND '182	USING 'LS181 AND '182	USING 'S181 AND 'S182	ARITHMETIC/ LOGIC UNITS	LOOK-AHEAD CARRY GENERATORS	
1 to 4	24 ns	24 ns	11 ns	1		NONE
5 to 8	36 ns	40 ns	18 ns	2		RIPPLE
9 to 16	36 ns	44 ns	19 ns	3 or 4	1	FULL LOOK-AHEAD
17 to 64	60 ns	68 ns	28 ns	5 to 16	2 to 5	FULL LOOK-AHEAD

description

The '181, 'LS181, and 'S181 are arithmetic logic units (ALU)/function generators which have a complexity of 75 equivalent gates on a monolithic chip. These circuits perform 16 binary arithmetic operations on two 4-bit words as shown in Tables 1 and 2. These operations are selected by the four function-select lines (S0, S1, S2, S3) and include addition, subtraction, decrement, and straight transfer. When performing arithmetic manipulations, the internal carries must be enabled by applying a low-level voltage to the mode control input (M). A full carry look-ahead scheme is made available in these devices for fast, simultaneous carry generation by means of two cascade-outputs (pins 15 and 17) for the four bits in the package. When used in conjunction with the SN54182, SN54S182, SN74182, or SN74S182, full carry look-ahead circuits, high-speed arithmetic operations can be performed. The typical addition times shown above

Figure 13.2 74181 Arithmetic and Logic Unit.

The signal designations shown below result in the logic functions and arithmetic operations shown in the truth table.

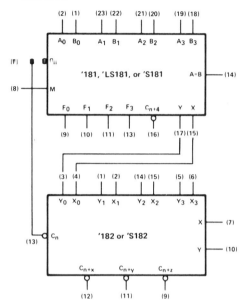

SELECTION					ACTIVE-HIGH DATA		
					M = H	M = L; ARITHMETIC OPERATIONS	
				LOGIC		C_n = H	C_n = L
S3	S2	S1	S0	FUNCTIONS		(no carry)	(with carry)
L	L	L	L	$F = \overline{A}$	F = A	F = A PLUS 1
L	L	L	H	$F = \overline{A + B}$	F = A + B	F = (A + B) PLUS 1
L	L	H	L	$F = \overline{A}B$	$F = A + \overline{B}$	$F = (A + \overline{B})$ PLUS 1
L	L	H	H	F = 0	F = MINUS 1 (2's COMPL)	F = ZERO
L	H	L	L	$F = \overline{AB}$	$F = A$ PLUS $A\overline{B}$	$F = A$ PLUS $A\overline{B}$ PLUS 1
L	H	L	H	$F = \overline{B}$	$F = (A + B)$ PLUS $A\overline{B}$	$F = (A + B)$ PLUS $A\overline{B}$ PLUS 1
L	H	H	L	$F = A \oplus B$	F = A MINUS B MINUS 1	F = A MINUS B
L	H	H	H	$F = A\overline{B}$	$F = A\overline{B}$ MINUS 1	$F = A\overline{B}$
H	L	L	L	$F = \overline{A} + B$	F = A PLUS AB	F = A PLUS AB PLUS 1
H	L	L	H	$F = \overline{A \oplus B}$	F = A PLUS B	F = A PLUS B PLUS 1
H	L	H	L	F = B	$F = (A + \overline{B})$ PLUS AB	$F = (A + \overline{B})$ PLUS AB PLUS 1
H	L	H	H	F = AB	F = AB MINUS 1	F = AB
H	H	L	L	F = 1	F = A PLUS A*	F = A PLUS A PLUS 1
H	H	L	H	$F = A + \overline{B}$	F = (A + B) PLUS A	F = (A + B) PLUS A PLUS 1
H	H	H	L	F = A + B	$F = (A + \overline{B})$ PLUS A	$F = (A + \overline{B})$ PLUS A PLUS 1
H	H	H	H	F = A	F = A MINUS 1	F = A

*Each bit is shifted to the next more significant position.

illustrate the little additional time required for addition of longer words when full carry look-ahead is employed. The method of cascading '182 or 'S182 circuits with these ALU's to provide multi-level full carry look ahead is illustrated under typical applications data for the '182 and 'S182.

Subtraction is accomplished by 1's complement addition where the 1's complement of the subtrahend is generated internally. The resultant output is A−B−1 which requires an end-around or forced carry to provide A−B.

The '181, 'LS181 or 'S181 can also be utilized as a comparator. The A = B output is internally decoded from the function outputs (F0, F1, F2, F3) so that when two words of equal magnitude are applied at the A and B inputs, it will assume a high level to indicate equality (A = B). The ALU should be in the subtract mode with C_n = H when performing this comparison. The A = B output is open-collector so that is can be wire-AND connected to give a comparison for more than four bits. The carry output (C_{n+4}) can also be used to supply relative magnitude information. Again, the ALU should be placed in the subtract mode by placing the function select inputs S3, S2, S1, S0 at L, H, H, L, respectively.

INPUT C_n	OUTPUT C_{n+4}	ACTIVE-HIGH DATA (FIGURE 1)	ACTIVE-LOW DATA (FIGURE 2)
H	H	A ≤ B	A ≥ B
H	L	A > B	A < B
L	H	A < B	A > B
L	L	A ≥ B	A ≤ B

These circuits have been designed to not only incorporate all of the designer's requirements for arithmetic operations, but also to provide 16 possible functions of two Boolean variables without the use of external circuitry. These logic functions are selected by use of the four function-select inputs (S0, S1, S2, S3) with the mode-control input (M) at a high level to disable the internal carry. The 16 logic functions are detailed in Tables 1 and 2 and include exclusive-OR, NAND, AND, NOR, and OR functions.

TEXAS INSTRUMENTS
INCORPORATED

Figure 13.2 Continued.

EXPERIMENT 14: DIGITAL TO ANALOG CONVERSION

A. BACKGROUND

This experiment makes use of a multiplying D/A, which functions as a digital "volume control." The flexibility of the multiplying function is useful in illustrating the widest range of D/A applications, particularly in instrumentation. This device uses CMOS technology, and is consequently more easily destroyed than a TTL device. Follow the precautionary notes below carefully.

B. REFERENCES

Text, Chapter 9. See also Analog Devices Databook.

C. PROBLEMS

Text, Problems 9.1, 9.2, and 9.3.

D. EQUIPMENT

Multiplying D/A converter (**AD 7520** or **7533**)
op amp (a 741 will do for audio or dc applications)
$\pm$15 V and +5 V **power supplies**
TTL IC's
    7493 and 74193 counters
    7432 and 7408 gates or equivalent
    7474 inverter (or gate equivalent)
555 wired as a variable frequency clock (subaudio to high audio is adequate)
    7470 JK edge-triggered flip flop or equivalent
    7489 memory
Outboards: clock, two 4-bit switch registers, two 4-bit LED lights, pulser.

DIGITAL TO ANALOG   14-1

Scope and camera

Audio amplifier and speaker (desirable, though not entirely necessary)

CMOS PRECAUTIONS

CMOS devices are more fragile than TTL devices, and can be destroyed by static charge buildup, overvoltages, or improper application of power. Many of these problems are becoming less important due to internal protection devices, but the following notes are a useful/safe starting point for the first use of a CMOS IC.

1. Leave the IC in its black conducting tray until ready to use. When handling it, ground yourself or at least touch a ground just before handling the IC. Any soldering iron must be grounded if it is used while the IC is in the circuit. When inserting a CMOS IC into a socket, a test clip whose pins are connected together to ground is the safest way to handle the IC.

2. To protect against power supply problems:

Use diodes in series with V_{dd} to guard against the certain destruction which will occur if polarities are reversed. Remove them after debugging the circuit, since they may interfere with the D/A accuracy.

Sequence power supply and reference turn-on so that the digital power is sure to be on whenever a reference appears (V_{dd} first on, last off). This ensures against a runaway of the FET switches.

Do not apply voltages higher than V_{dd} or less than ground on any terminal except V_{ref}.

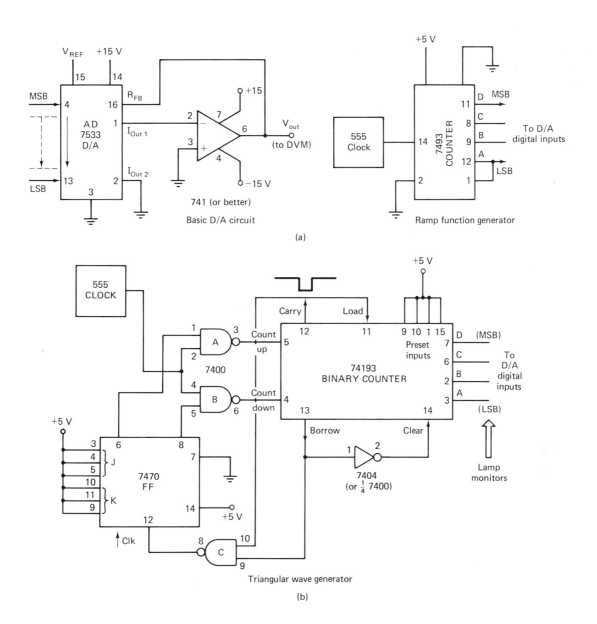

Figure 14.1 (a) Basic D/A circuit and D/A function generator for ramp waveform. (b) D/A function generator for triangular waveform.

E. MEASUREMENTS

1. Basic testing of the digital analog function. Wire up a basic D/A circuit, as shown in Fig. 14.1(a). Use 8 bits of switch register as inputs, and a DVM as output. With a 10-bit D/A, the unused bits can be taken care of by wiring the two most significant bits to ground. Use the op amp +15 V power supply as the source of V_{ref} and V_{dd} (though this can be varied if desired). Use a 5V TTL power supply to drive the switch register inputs.

Vary the switch register inputs and verify that the analog output follows the digital input. Check the full scale (all 1's) and the smallest number (all 0's). Verify initially that the D/A is functioning by checking one bit at a time. By comparing the observed outputs with that expected from the transfer function, check whether the output has the accuracy expected (8 bits = what percentage uncertainty?). Record a series of input/output measurements and check the linearity of the device.*

2. Function Generation: triangles and ramps.

a. Wire up a 7493 counter as input, driven by variable frequency clock, as shown in the right side of Fig. 14.1(a). Wire unused D/A inputs to ground. Observe the output on a scope when the input is at audio frequency. Is the observed waveform what you expected? Now increase the input frequency to past 100 kHz. Look at the change of output state carefully. Can you see any delay due to the settling time of D/A, or op amp slew rate? Do you observe any **glitches**, sharp spikes at the point of transition? What might they be due to?

(Optional) With the input at audio frequencies, connect the output to an amplifier and speaker. Can you hear the difference from a conventional ramp "sound" generated by a conventional function generator? What is the lowest harmonic at which the D/A approximation differs from a true ramp?

---

* There are two ways: equally spaced inputs and a linear plot; or, one bit at a time input and a log plot, since $2^n = \exp[n \ln 2]$. In both cases, relate the deviation to $\pm 1/2$ LSB.

b. It is possible to generate a triangular wave digitally, using a 74193 counter with its up/down capability. Wire up the circuit shown in Fig. 14.1(b).

(You may wish to see what happens when the circuit is modified from that shown. In particular, what happens when the load or clear inputs are disconnected?) Use LED readout lights on the counter output to initially test circuit operation at low (1 Hz) frequency.

With the input at audio frequency, observe the output on a scope, and compare it with what you expect. How many steps are observed? Raise the input frequency to 100 kHz and look for glitches at switching transitions. Compare with that observed for the 7493, and explain any difference. (Recall: the 7493 is a ripple counter, but the 74193 is synchronous.)

(Optional) Connect an audio amplifier and speaker, and repeat the observations you made with a ramp signal input.

3. Arbitrary Functions, using memory. (Optional, but fun; do Exp. 11 first.) Replace the 7493 counter with a 7489 16 4-bit word memory, as shown in Fig. 14.2. (Review experiment 11 if needed.) Load a series of binary words to generate some function. A suggestion is a sine wave, **offset** so the **minimum** and **maximum** are stored as **0000** and **1111**.

Work out the closest 4-bit approximation to the function you choose, and load it in memory. Cycle the memory address through a complete 16-word cycle, and observe the digital output with LED lights, and the corresponding analog output with a DVM. Now clock the memory address at an audio frequency, and observe the output on a scope. Does the waveform agree with what you tried to put in digital form?

(Optional) Listen to the audio signal with amplifier and speaker. Compare with conventional audio oscillator signal.

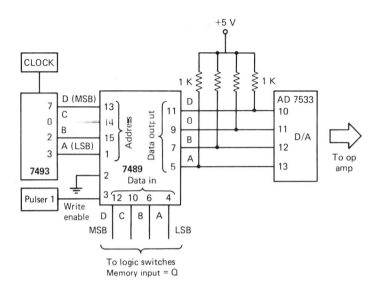

Figure 14.2 D/A function generator. Arbitrary waveform using RAM or ROM. Since the D/A analog is wired single-ended, only waveforms positive with respect to ground will appear at the output.

4. Multiplying D/A operation. The input V_{ref} need not be a constant, but can be an arbitrary test function. This allows digital programming of analog test instruments. Try it for a sine wave from an audio oscillator input at V_{ref}. (Make sure that the amplitude does not exceed the D/A's specification.) Use an 8-bit switch register as the digital input, and verify that you can digitally set the sine wave output to any with 8-bit precision.

5. (Optional) Digital Electronic Music. The D/A is the basic interface from the digital to the analog world. Given a function in digital form and a variable frequency clock, basic digitally generated music can be explored. Use any of the previous digital input functions. Connect the output to an amplifier and speaker. The input frequency can be varied in a conventional way. More interesting is the possibility of approximating the twelve tone scale with single

14-6   DIGITAL TO ANALOG

keystrokes, using a programmable divide-by-N counter, or with a top-active synthesizer. (See Text, Fig. 5.14.) Another optional extension of this section is to explore the consequences of having **two** arbitrary functions, one input at V_{ref} and the other at the digital input. Both can have independent frequency and form. The multiplying D/A output frequency spectrum is similar to that of an **analog** multiplier, (see Experiment 22), or of a music synthesizer. But since one of the inputs here is digital, the range of possible output signals is limited only by the imagination.

GENERAL DESCRIPTION

The AD7533 is a low cost 10-bit 4-quadrant multiplying DAC manufactured using an advanced thin-film-on-monolithic-CMOS wafer fabrication process.

Pin and function equivalent to the industry standard AD7520, the AD7533 is recommended as a lower cost alternative for old AD7520 sockets or new 10-bit DAC designs.

AD7533 application flexibility is demonstrated by its ability to interface to TTL or CMOS, operate on +5V to +15V power, and provide proper binary scaling for reference inputs of either positive or negative polarity.

PRELIMINARY TECHNICAL DATA

FEATURES
Lowest Cost 10-Bit DAC
Direct AD7520 Equivalent
Linearity: ½, 1 or 2LSB
Low Power Dissipation
Full Four-Quadrant Multiplying DAC
CMOS/TTL Direct Interface

APPLICATIONS
Digitally Controlled Attenuators
Programmable Gain Amplifiers
Function Generation
Linear Automatic Gain Control

FUNCTIONAL DIAGRAM

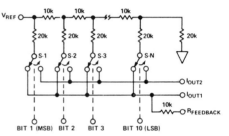

DIGITAL INPUTS (DTL/TTL/CMOS COMPATIBLE)

Logic: A switch is closed to $I_{OUT\,1}$ for its digital input in a "HIGH" state.

ORDERING INFORMATION

Nonlinearity	Temperature Range and Package		
	Commercial (Plastic) 0 to +70°C	Industrial (Ceramic) −25°C to +85°C	Military (Ceramic) −55°C to +125°C
±0.2%	AD7533JN	AD7533AD AD7533AD/883B	AD7533SD AD7533SD/883B
±0.1%	AD7533KN	AD7533BD AD7533BD/883B	AD7533TD AD7533TD/883B
±0.05%	AD7533LN	AD7533CD AD7533CD/883B	AD7533UD AD7533UD/883B

PIN CONFIGURATION

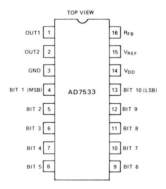

Figure 14.3 7533 Specification Sheets.

SPECIFICATIONS

(V_{DD} = +15V; V_{OUT1} = V_{OUT2} = 0V; V_{REF} = +10V unless otherwise noted)

PARAMETER	T_A = 25°C	T_A = Operating Range[1]	Test Conditions
STATIC ACCURACY			
Resolution	10 Bits	10 Bits	
Nonlinearity[2]			
AD7533JN, AD, SD	±0.2% FSR max	±0.2% FSR max	
AD7533KN, BD, TD	±0.1% FSR max	±0.1% FSR max	
AD7533LN, CD, UD	±0.05% FSR max	±0.05% FSR max	
Gain Error[3]	±1.4% FS max	±1.5% FS max	Digital Inputs = V_{INH}
Supply Rejection[4]			
ΔGain/ΔV_{DD}	0.005%/%	0.008%/%	Digital Inputs = V_{INH}; V_{DD} = +14V to +17V
Output Leakage Current			
I_{OUT1} (pin 1)	±50nA max	±200nA max	Digital Inputs = V_{INL}; V_{REF} = ±10V
I_{OUT2} (pin 2)	±50nA max	±200nA max	Digital Inputs = V_{INH}; V_{REF} = ±10V
DYNAMIC ACCURACY			
Output Current Settling Time	600ns max[5]	800ns[4]	To 0.05% FSR; R_{LOAD} = 100Ω; Digital Inputs = V_{INL} to V_{INH} or V_{INL} to V_{INH}
Feedthrough Error	±0.05% FSR max[4]	±0.1% FSR max[4]	Digital Inputs = V_{INL}; V_{REF} = ±10V, 100kHz sinewave.
REFERENCE INPUT			
Input Resistance (pin 15)	5kΩ min, 20kΩ max	5kΩ min, 20kΩ max[6]	
ANALOG OUTPUTS			
Output Capacitance			
C_{OUT1} (pin 1)	100pF max[4]	100pF max[4]	Digital Inputs = V_{INH}
C_{OUT2} (pin 2)	35pF max[4]	35pF max[4]	
C_{OUT1} (pin 1)	35pF max[4]	35pF max[4]	Digital Inputs = V_{INL}
C_{OUT2} (pin 2)	100pF max[4]	100pF max[4]	
DIGITAL INPUTS			
Input High Voltage			
V_{INH}	2.4V min	2.4V min	
Input Low Voltage			
V_{INL}	0.8V max	0.8V max	
Input Leakage Current			
I_{IN}	±1µA max	±1µA max	V_{IN} = 0V and V_{DD}
Input Capacitance			
C_{IN}	5pF max[4]	5pF max[4]	
POWER REQUIREMENTS			
V_{DD}	+15V ±10%	+15V ±10%	Rated Accuracy
V_{DD} Range[4]	+5V to +16V	+5V to +16V	Functionality with degraded performance
I_{DD}	2mA max	2mA max	Digital Inputs = V_{INL} or V_{INH}
I_{DD}	100µA max	150µA max	Digital Inputs = 0V or V_{DD}

NOTES:
[1] Plastic (JN, KN, LN versions): 0 to +70°C
Commercial Ceramic (AD, BD, CD versions): -25°C to +85°C
Military Ceramic (SD, TD, UD versions): -55°C to +125°C
[2] "FSR" is Full Scale Range.
[3] Full Scale (FS) = $-(V_{REF})\left(\frac{1023}{1024}\right)$
[4] Guaranteed, not tested.
[5] AC parameter, sample tested to ensure specification compliance.
[6] Absolute temperature coefficient is approximately -300ppm/°C.
[7] 100% screened to MIL-STD-883, method 5004, para. 3.1.1 through 3.1.12 for class B device. Final electrical tests are: Nonlinearity, Gain Error, Output Leakage Current, V_{INH}, V_{INL}, I_{IN} and I_{DD} at +25°C and +125°C (SD, TD, UD versions) or +25°C and +85°C (AD, BD, CD versions).

Specifications subject to change without notice.

Figure 14.3 Continued.

ABSOLUTE MAXIMUM RATINGS
(T_A = +25°C unless otherwise noted)

V_{DD} to GND	−0.3V, +17V
R_{FB} to GND	±25V
V_{REF} to GND	±25V
Digital Input Voltage Range	−0.3V to V_{DD}
Output Voltage (pin 1, pin 2)	−0.3V to V_{DD}
Power Dissipation (Package)	
Plastic (Suffix N)	
To +70°C	670mW
Derates above +70°C by	8.3mW/°C
Ceramic (Suffix D)	
To +75°C	450mW
Derates above +75°C by	6mW/°C
Operating Temperature Range	
Commercial (JN, KN, LN versions)	0 to +70°C
Industrial (AD, BD, CD versions)	−25°C to +85°C
Military (SD, TD, UD versions)	−55°C to +125°C
Storage Temperature	−65°C to +150°C
Lead Temperature (Soldering, 10 seconds)	+300°C

CAUTION:

1. ESD sensitive device. The digital control inputs are Zener protected; however, permanent damage may occur on unconnected devices subjected to high energy electrostatic fields. Unused devices must be stored in conductive foam or shunts.
2. Do not apply voltages lower than ground or higher the V_{DD} to any pin except V_{REF} (pin 15) and R_{FB} (pin 16).
3. The inputs of some IC amplifiers (especially wide bandwidth types) present a low impedance to V⁻ during power-up or power-down sequencing. To prevent the AD7533 OUT1 or OUT2 terminals from exceeding −300mV (which causes catastrophic substrate current) a Schottky diode (HP5082-2811 or equivalent) is recommended. The diode should be connected between OUT1 (OUT2) and ground as shown in Figures 5 and 6.

TERMINOLOGY

NONLINEARITY: Error contributed by deviation of the DAC transfer function from a best straight line function. Normally expressed as a percentage of full scale range. For a multiplying DAC, this should hold true over the entire V_{REF} range.

RESOLUTION: Value of the LSB. For example, a unipolar converter with n bits has a resolution of (2^{-n}) (V_{REF}). A bipolar converter of n bits has a resolution of $[2^{-(n-1)}]$ [V_{REF}]. Resolution in no way implies linearity.

SETTLING TIME: Time required for the output function of the DAC to settle to within 1/2 LSB for a given digital input stimulus, i.e., 0 to Full Scale.

GAIN: Ratio of the DAC's operational amplifier output voltage to the input voltage.

FEEDTHROUGH ERROR: Error caused by capacitive coupling from V_{REF} to output with all switches OFF.

OUTPUT CAPACITANCE: Capacity from I_{OUT1} and I_{OUT2} terminals to ground.

OUTPUT LEAKAGE CURRENT: Current which appears on I_{OUT1} terminal with all digital inputs LOW or on I_{OUT2} terminal when all inputs are HIGH.

Figure 14.3 Continued.

CIRCUIT DESCRIPTION

GENERAL CIRCUIT INFORMATION

The AD7533, a 10-bit multiplying D/A converter, consists of a highly stable thin film R-2R ladder and ten CMOS current switches on a monolithic chip. Most applications require the addition of only an output operational amplifier and a voltage or current reference.

The simplified D/A circuit is shown in Figure 1. An inverted R-2R ladder structure is used — that is, the binarily weighted currents are switched between the I_{OUT1} and I_{OUT2} bus lines, thus maintaining a constant current in each ladder leg independent of the switch state.

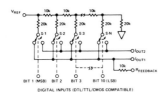

Figure 1. AD7533 Functional Diagram

One of the CMOS current switches is shown in Figure 2. The geometries of devices 1, 2 and 3 are optimized to make the digital control inputs DTL/TTL/CMOS compatible over the full military temperature range. The input stage drives two inverters (devices 4, 5, 6 and 7) which in turn drive the two output N-channels. The "ON" resistances of the switches are binarily scaled so the voltage drop across each switch is the same. For example, switch 1 of Figure 2 was designed for an "ON" resistance of 20 ohms, switch 2 or 40 ohms and so on. For a 10V reference input, the current through switch 1 is 0.5mA, the current through switch 2 is 0.25mA, and so on, thus maintaining a constant 10mV drop across each switch. It is essential that each switch voltage drop be equal if the binarily weighted current division property of the ladder is to be maintained.

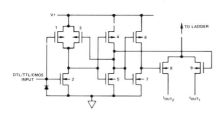

Figure 2. CMOS Switch

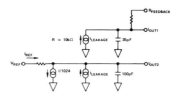

Figure 3. AD7533 Equivalent Circuit — All Digital Inputs Low

EQUIVALENT CIRCUIT ANALYSIS

The equivalent circuits for all digital inputs high and all digital inputs low are shown in Figures 3 and 4. In Figure 3 with all digital inputs low, the reference current is switched to I_{OUT2}. The current source $I_{LEAKAGE}$ is composed of surface and junction leakages to the substrate while the $\frac{1}{1024}$ current source represents a constant 1-bit current drain through the termination resistor on the R-2R ladder. The "ON" capacitance of the output N channel switch is 120pF, as shown on the I_{OUT2} terminal. The "OFF" switch capacitance is 30pF, as shown on the I_{OUT1} terminal. Analysis of the circuit for all digital inputs high, as shown in Figure 4, is similar to Figure 3; however, the "ON" switches are now on terminal I_{OUT1}, hence the 100pF at that terminal.

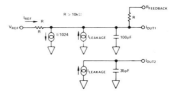

Figure 4. AD7533 Equivalent Circuit — All Digital Inputs High

Figure 14.3 Continued.

DIGITAL TO ANALOG   14-11

OPERATION
UNIPOLAR BINARY OPERATION (2-QUADRANT MULTIPLICATION)

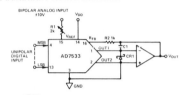

Figure 5. Unipolar Binary Operation (2-Quadrant Multiplication)

DIGITAL INPUT		NOMINAL ANALOG OUTPUT (V_{OUT} as shown in Figure 1)
MSB	LSB	
1111111111		$-V_{REF}\left(\frac{1023}{1024}\right)$
1000000001		$-V_{REF}\left(\frac{513}{1024}\right)$
1000000000		$-V_{REF}\left(\frac{512}{1024}\right) = -\frac{V_{REF}}{2}$
0111111111		$-V_{REF}\left(\frac{511}{1024}\right)$
0000000001		$-V_{REF}\left(\frac{1}{1024}\right)$
0000000000		$-V_{REF}\left(\frac{0}{1024}\right) = 0$

NOTES:
1. Nominal Full Scale for the circuit of Figure 5 is given by FS = $-V_{REF}\left(\frac{1023}{1024}\right)$
2. Nominal LSB magnitude for the circuit of Figure 5 is given by LSB = $V_{REF}\left(\frac{1}{1024}\right)$

Table 1. Unipolar Binary Code Table

BIPOLAR OPERATION (4-QUADRANT MULTIPLICATION)

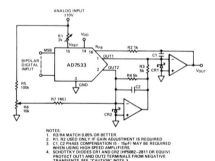

Figure 6. Bipolar Operation (4-Quadrant Multiplication)

DIGITAL INPUT		NOMINAL ANALOG OUTPUT (V_{OUT} as shown in Figure 2)
MSB	LSB	
1111111111		$-V_{REF}\left(\frac{511}{512}\right)$
1000000001		$-V_{REF}\left(\frac{1}{512}\right)$
1000000000		0
0111111111		$+V_{REF}\left(\frac{1}{512}\right)$
0000000001		$+V_{REF}\left(\frac{511}{512}\right)$
0000000000		$+V_{REF}\left(\frac{512}{512}\right)$

NOTES:
1. Nominal Full Scale Range for the circuit of Figure 6 is given by FSR = $V_{REF}\left(\frac{1023}{512}\right)$
2. Nominal LSB magnitude for the circuit of Figure 6 is given by LSB = $V_{REF}\left(\frac{1}{512}\right)$

Table 2. Bipolar (Offset Binary) Code Table

APPLICATIONS

10-BIT AND SIGN MULTIPLYING DAC

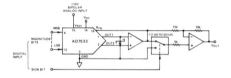

PROGRAMMABLE FUNCTION GENERATOR

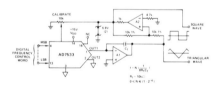

DIGITALLY PROGRAMMABLE LIMIT DETECTOR

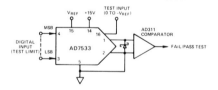

Figure 14.3 Continued.

EXPERIMENT 15:

ANALOG TO DIGITAL CONVERSION BY THE DUAL SLOPE METHOD

A. BACKGROUND

The dual slope method of A to D conversion has become the most popular one for use in Digital Voltmeters, where the demand is for high resolution and high accuracy, but only moderate speed. For example,

Resolution: 3 to 6 decimal digits
Accuracy: 0.1% to 0.005%
Speed: 1 s to 0.01 s per point

The general idea of dual slope operation is shown in the text, Fig. 9.14. In this experiment you will work with one version of a dual slope A/D, shown in Fig. 15.1. It is recommended that the circuit be prewired with test points brought out, so that you can quickly explore the operation of various portions of the circuit.

The dual slope design has excellent built-in noise immunity due to two features:

1. Since an integrator is present, high frequency noise is attenuated, since the integrator gain falls off as $1/f$.

2. Noise whose period is at multiples of the sampling frequency is <u>ideally nulled</u> (in practice 50-100 dB is obtained). Since one of the largest noise sources is ac pickup of harmonics from the power line, it is desirable to derive the sampling period from the ac line frequency. An example of the integrator output in the presence of (artificially generated) noise is shown in the Text, Fig. 9.16. Note that the "integral" (final integrator output at the end of time T_1) is relatively unchanged by the noise.

ANALOG TO DIGITAL 15-1

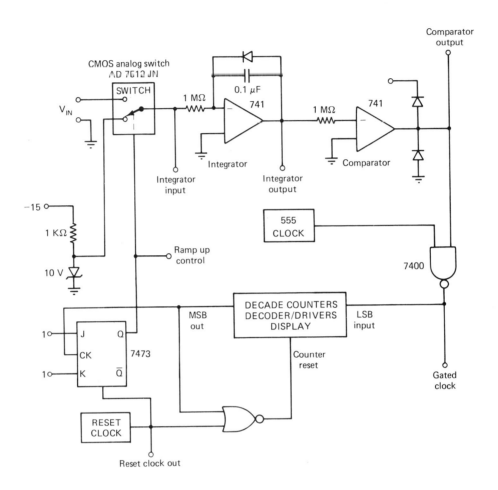

Figure 15.1 Practical working dual slope A/D test circuit. Note the labeled test points for examining circuit operation.

B. REFERENCES

Text, Chapter 9 Section 5.
Analog Devices A-D Conversion Handbook, p. II-48 to II-50.

15-2   ANALOG TO DIGITAL

C. PROBLEMS
Text, Problems 9.5, 9.6, and 9.7.

D. EQUIPMENT
Dual-slope A/D Module
Oscilloscope
Variable voltage source

E. MEASUREMENTS

1. Connect a variable dc input to the circuit. Using a dual-trace scope, examine and photograph the waveforms at the points shown in text, Problem 9.6. Trigger the oscilloscope on the comparator output, with the triggering mode selected so that a triangular up/down waveform is seen from the integrator output. How does the waveform change when the input voltage is increased? Compare your results with what you expected based on Prob. 9.6, and explain any unexpected observations.

2. With the same variable DC input, record the digitized output for a series of input voltages ranging from 0.0 V to full-scale on the counter. Plot the results, and examine the Linearity, Accuracy, Precision, Zero Offset, and Stability of the circuit.

3. Explore the noise immunity of the dual slope design. Sum* in with the dc voltage a variable-frequency sine wave source, of roughly 1/5 the amplitude of the DC input. Examine the waveforms at the integrator input and output, and note changes in the counter output as a result of this "noise source." Vary the frequency of the source, and examine how the digitized output error decreases as f increases. Look especially for **nulls** in the error at multiples of the sampling frequency.

* Consult your instructor on how to do this. Note to instructor: An Op-Amp summing circuit is best, though passive summing will work if you give the dc and ac signals a source of impedance the same magnitude (try 1K Ohm).

EXPERIMENT 16:

ANALOG TO DIGITAL CONVERSION

BY THE SUCCESSIVE APPROXIMATION METHOD

A. BACKGROUND

The successive approximation method is the most common way to encode in binary form (= <u>digitize</u>) an analog signal which is varying too fast to be seen by a dual slope digital voltmeter. Two versions are given. The first requires only standard TTL IC's for the logic, but is relatively complex to wire. The second version has all of the successive approximation logic on a single IC, and is far preferable if the IC is available.

Finally, an optional but extremely interesting section shows you how to display on a scope the point of the waveform which is being digitized.

B. REFERENCES
   None.

C. PROBLEMS
   None.

D. EQUIPMENT
   1 D/A (AD 7520 or AD 7533).
   IC's as shown in Fig. 16.1 or Fig. 16.2 (USE THE SELF-CONTAINED SUCCESSIVE APPOX. CHIP IF AVAILABLE, TO SAVE WIRING).
   scope
   digital voltmeter
   input source: battery and voltage divider or more interesting input.

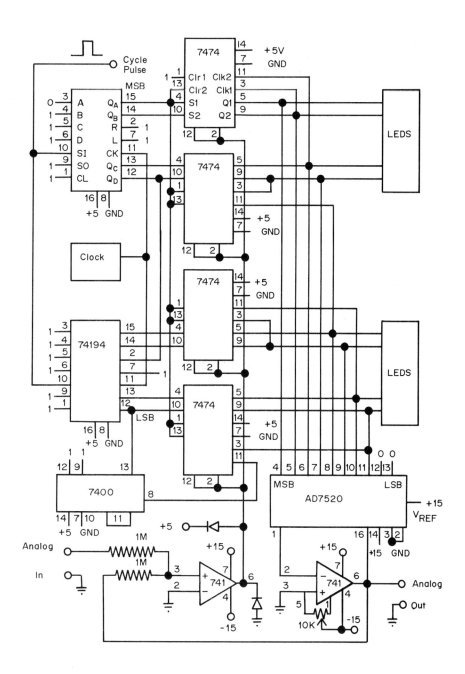

Figure. 16.1 Wiring diagram for an 8-bit successive-approximation A/D from TTL MSI IC's. This version is to be avoided if the self-contained successive approximation logic IC of Fig. 16.2 is available, since it is painful to wire and debug. This circuit is due to T. G. Matheson.

16-2   SUCCESSIVE APPROXIMATION METHOD

Circuit notes (Fig. 16.1 version)

1. The load or cycle pulse (74194 pin 9) must be high on the rising edge of the clock.

2. Control high loads 0111 .... 11 into the S-R.

3. Control low shifts right and inserts 1's on every rising clock edge:

   First rising clock edge after loading:
   $$10111 ... 11$$
   2nd edge: $11011 ... 11$
   etc.

After the 0 is shifted out, the cycle is complete, as the S-R continues shifting:
$$11111 ... 111$$
$$11111 ... 111$$
etc.

4. The op-amp output range (~±12 V) limits D/A converter.

5. Presetting the MSB <u>resets</u> the rest of the D FF/s.

6. Each D FF is preset by a "0" from S-R, independent of the clock.

7. Each D FF is clocked by the presetting of the following D. The last (LSB) D FF is clocked by a time delayed rising edge from the LSB output of the S-R.

8. The op-amp is nulled initially with 000 ... 00 input to the D/A.

Construction Notes (Fig. 16.1 version). 2 SK-10's are needed.

1. Wire the 74194 SR first and test with the LED's.

2. Add the 7474 D-FF's and test their output using a very slow clock or by hand and a switch to insert "data" (replacing comparator).

3. Wire up the circuit board before inserting the D/A. Add the D/A. The CMOS IC is safe against static once in a wired SK 10.

4. Turn the CMOS ±15 power supply **ON FIRST and OFF LAST**.

5. Test through to the output of the comparator before completing the connection of comparator output to data input.

6. **Don't** try to protect the D/A with diodes on the power supply. It works-- but with 10% non-linearity. Instead, just don't make any mistakes in wiring in the power supply voltages!

E. MEASUREMENTS

1. Tests on the D/A converter. This is not necessary if you have previously done experiment 13, although it might be best to repeat it to ensure that the D/A being used has the desired linearity. First, turn on each of the 8 data bits individually and measure the output voltage with a digital voltmeter. The results should fit on an exponential curve. Second, set a range of digital inputs spanning a roughly linear range of output voltages. In both cases, the deviation from linearity can be compared with the expected D/A deviation of ± 1/2 LSB.

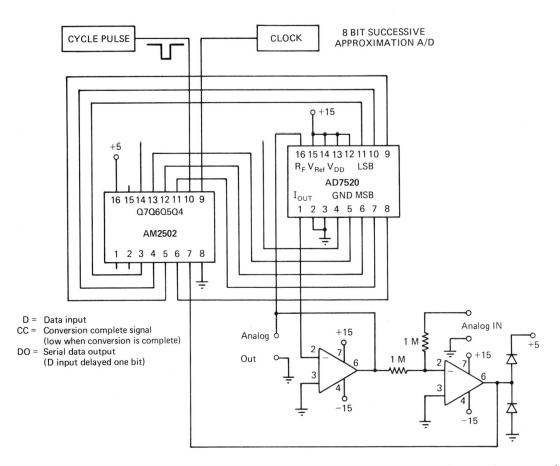

Fig. 16.2 Three-chip successive approximation A/D converter. The AM 2502 performs all of the logic sequences of the previous circuit.

2. Tests on the complete A/D circuit. Measure the digital output and the analog output as a function of input voltage over a linear range from 0 V to full-scale. Plot the digital output as a function of analog input, and check the linearity. A useful form is to take the decimal equivalent of the binary output and measure the deviation in those arbitrary digital units. Plot the ratio of analog output to analog input, as a function of analog input. The results should be a straight line whose slope is 1.0, with a standard deviation related to the LSB of the D/A. The simplest input is a battery whose value is close to the full scale range, and a variable resistor hooked up as a voltage divider.

SUCCESSIVE APPROXIMATION METHOD   16-5

3. More interesting inputs. The most important applications of this circuit come in grabbing a signal which is too fast or too rapidly changing to be seen with a conventional digital voltmeter. For example, record the amplitude at a certain point during a pulse. The beginning of the pulse itself is used to drive the encode command. To generate a simple test example, run a square wave through a high pass filter to generate an exponentially damped pulse. Design a way to accurately measure the level of any point on that pulse. A one-shot is needed to delay the encode command to the desired part of the waveform. By varying the delay (variable resistor in the RC), one can map out a picture of the waveform with much higher accuracy than an oscilloscope is capable of.

OPTIONAL (Consult instructor). To get a visual display on a scope of where the digital sample is being taken, picture on a scope, one can derive a pulse from the encode command which then goes to the scope Z-axis, making a bright spot on the waveform at the point currently being digitized.

EXPERIMENT 17:

BASIC OPERATIONAL AMPLIFIER CIRCUITS

A. BACKGROUND

The operational amplifier allows someone who doesn't know very much electronics to do subtle manipulations with signals. The basic operational amplifier circuit is shown in Text, Fig. 11.8. With large loop gain **a** and negative feedback, the circuit gain depends only on the value of passive components in the input and feedback branches:

$$\frac{V_o}{V_i} \cong \frac{-Z_f}{Z_i} \quad \text{for} \quad \frac{Z_i}{Z_f} \; a \gg 1$$

With reactive circuit elements, the circuit can act as a time integrator or time differentiator. In understanding these and other operational amplifier applications, it is useful (and very nearly correct) to imagine that the operational amplifier senses the voltage difference across its inputs and causes an output V_o to appear which is just sufficient to drive the point **S** in Fig. 11.8 to ground potential via the feedback element Z_f.

B. REFERENCES

Text, Chapter 11, and portions of Chapter 12, as needed.

C. PROBLEMS

Chapter 11, Problems 1, 2, 5, 8, and 9 are minimum preparation for this experiment.

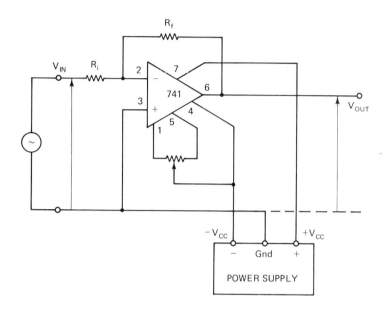

Figure 17.1 Offset nulling circuit for 741 op amp. Rn, the offset trim adjust, has a typical value of 10 K Ohms.

D. EQUIPMENT

1. Op Amps. As op amps, we will be using the 741, a moderately high quality but low-cost IC in an 8-pin dual-inline (DIP) package. Refer to Fig. 17.1 for physical pin layout and specifications. One nice thing about the DIP package is that there are only two ways to insert it. But since a mistake will apply V+ to the input, and destroy the op amp, be careful. The clue is a mark on the package on the end nearest pins 1 and 8.

You will discover that a high gain op amp is sensitive to pickup of external noise. Keep leads short, and when long leads are necessary (e.g., to power supply or scope) use shielded cable or twisted pair.

<u>Precautions</u> <u>with</u> <u>741's</u> <u>and</u> <u>any</u> <u>IC</u> <u>op</u> <u>amp.</u> There are four common causes of IC destruction:

a. There are maximum and minimum usable supply voltages. Below the minimum value, the transistors will not operate as linear amplifiers. When the maximum value is exceeded, the collector-base junctions break down, large currents flow, and the pn junctions diffuse away.

b. If a large input voltage (~5V) appears across the input terminals, there can be large transient currents flowing through the op amp, especially when there is a large feedback capacitor. Make sure the feedback loop is closed. (Why?)

c. Within the op amp, electronic isolation of the various parts depends upon reverse biased junctions. If the supply polarity is accidentally reversed, the junctions become forward biased, and large currents can flow, which can destroy the op amp.

d. A load that draws too much current can cause the op amp to heat up. Most operational amplifiers are now designed with internal current limiting (current clamps) to prevent this.

2. Breadboards: The Superstrip

Circuits will be built on the same breadboard used in the digital experiments. There are now two power supply voltages. Run the +15V along one edge, the -15 along the other, and color code the wires (red = plus) to avoid confusion. Run ground bus lines along the other two available bus strips. It is important to learn how to accurately wire up a circuit. Expect some frustration along the way, since linear analog circuits are more subtle than two valued digital circuits.

Useful Preparations

    a. A couple of coaxial cables with a BNC on one end and two solid wires at the other end are useful for connections to scope or other instruments.

    b. A dual tracking $\pm$ 15 V power supply is needed (~100 ma).

    c. Linear circuits will amplify noise. Pay careful attention to proper grounding, and use shielded cables for long connections to instruments.

    d. Variable resistors (pots) should be equipped with short, solid wire leads (2 1/2 to 3 inches). Even better are the trim pots with printed circuit board pins which plug into the breadboard.

    e. Capacitors placed between power supply pins and ground on the breadboard will eliminate noise caused by R-F pickup. Try 1 uf.

E. MEASUREMENTS

1. Balancing input-voltage offset. Connect a gain of 100 amplifier (Fig. 17.1) to amplify the voltage offset, using R_f = 100 K ohms and R_i = 1 K ohms. Ground the input V_i. The zero-offset adjustment is made with an external variable resistor as shown.

Turn the pot while observing V_o with a dc oscilloscope or digital voltmeter; a range of several volts on either side of zero should be obtainable. Zero the offset as well as possible, turning the scope or voltmeter eventually to its highest gain. If you are unable to balance the amplifier, you probably neglected a connection. Check the connections and, if still unsuccessful, call for help.

To get rid of the DC instability caused by RF pickup at the power supply pins on the 741, place 1 uf capacitors between each power supply pin <u>on the 741</u> and ground.

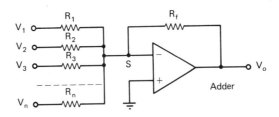

Figure 17.2 Adder

2. Adder. Replace R_i with two 100 K ohm resistors connected to the summing point to make an adder, as shown in Fig. 17.2. Using 1.3 volt mercury batteries or 1/10 of the power supply voltage (use a voltage divider), check the operation with a single input. Can you detect any voltage at the summing point? Check adder operation using two input voltages, first of the same sign and then (a real test) of opposite sign. Use the ±15 supply, since the voltages are nearly equal in magnitude. If you can detect any voltage V_s at the summing point, does its magnitude make sense? Recall that feedback reduces V_s to $V_i/(1 - \beta a)$. Now observe the adder operation using two sine wave generators, and note the best patterns possible.

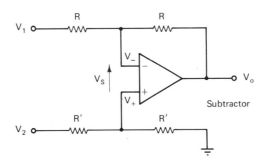

Figure 17.3 Subtractor

3. Subtractor. (Optional) Connect a subtractor voltage circuit as shown in Fig. 17.3. Observe its action for two arbitrary dc voltages. What voltage appears at each amplifier input? Can you detect any voltage difference between the inputs?

OPERATIONAL AMPLIFIERS

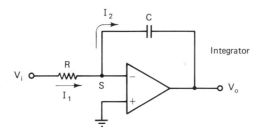

Figure 17.4 Integrator.

4. Integrator. A 741 makes a poor integrator at DC, but works fine at 1 HZ and above. Make sure you have the scope DC-coupled to avoid distorting the waveform, and reset the integrator (1 K ohm across C_f) when the output drifts off scale.

If you must, use a 741 as an integrator, by putting a large resistor (10 M ohms) across C_f so as to eliminate the dc drift.

Alternatively, it is recommended that you try a "super 741" op amp with much lower bias current, such as the BiMOS RCA CA3140. It is compatible with the 741, has bias current I_b 1000 times lower, and costs only about one dollar. A BiFET (TL080) is also a good solution.

Connect an integrator circuit as in Fig. 17.4, using a 1 uf capacitor for Z_f and a 100 K ohm resistor for Z_i. An integrator is initialized by shorting the capacitor to remove stored charge. Use a 1 K ohm resistor for this purpose to limit the current surge. When you remove this resistor, you will probably observe the output gradually drifting away from zero volts, even though the input voltage is Zero (short R_i to ground to guarantee this). Assuming zero capacitor leakage (polystyrene capacitors are used for this reason), the drift is due to non-ideal offset voltage and offset current in the op amp. The output voltage will tend to drift due to offset current I at a rate I/C, so that even if I is only 10^{-7} A and C = 1 uf, the output will drift to 1 V in only 10 seconds. Offset current is thus a prime consideration in choosing an operational amplifier for integrator applications. Offset voltage can also produce drift, driving a current through the feedback capacitor which creates output drift. Observe the rate of drift of your integrator. The drift is primarily due to offset current if you have carefully zeroed the voltage offset earlier. Estimate whether the offset current is within specifications for the amplifier used. Note that if offset voltage is adding to the drift, the value of R_i will affect the drift rate. Why?

Observe the integrator action for a variety of waveforms. Try a small constant voltage input and verify that the output ramp rate is governed by RC. Now observe the output <u>amplitude</u> with a sine-wave input as the frequency is varied. What is the relationship to the input amplitude? Check this over a wide frequency range. Finally, what does the output look like when the input is a square wave?

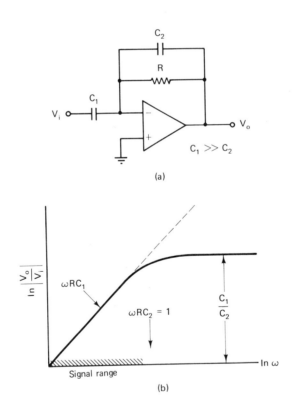

Figure 17.5 Practical Differentiator.

5. Differentiator. Connect a differentiator circuit (Fig. 17.5). Use R_f = 10 K ohm paralleled by 100 pF, and C_i = 0.1 uF. Apply a small sine-wave signal at about 100 Hz and make sure the amplifier isn't saturating. Using sine waves, verify that the circuit has the gain of a true differentiator, linear as a function of frequency. If the same RC components were used as a passive differentiator, at what frequency would it no longer differentiate?

Now put in a square wave; observe and describe the output. If the output spikes look non-ideal, don't neglect the possibility that this distortion is present in the input waveform. The oscilliscope probe must be properly compensated; see text, Appendix 2. Check that the circuit differentiates well from 10Hz to 1 KHz. Constant spike width is expected; is the width related to the input waveform?

Now try a triangular wave at some frequency. Observe the square wave character of the output, and try to modify the high-frequency gain to make the output nearly ideal without rounding the corners of the square wave.

Throughout the experiments, it is convenient to use a breadboard which allows quick circuit building without soldering, and also results in reliable connections. An example is shown in Fig. 2.6. The topology and pin spacing are particularly suited for work with IC's in the DIP package, an industry standard. Standard components (R, C, etc.) just plug in, and wiring is done by inserting a stripped solid wire (# 22 AWG is best).

The superstrip breadboard is cheap enough so that each student can retain control over one, attaching it to a base containing a power supply, etc., only when needed. This has a special value in saving time when rather complicated wiring is needed without needlessly tying up equipment. Be aware that learning how to accurately wire up a circuit is important, and expect some frustration along the way.

EXPERIMENT 18:

OPERATIONAL AMPLIFIER APPLICATIONS IN CONTROL:

THE VOLTAGE REGULATOR

A. BACKGROUND

In many experiments it is necessary to control physical variables so that they remain constant or vary in a programmed way. The variable may be a voltage, a current, a magnetic field, a light intensity, or an angular position. The negative feedback loop essential to such control is the same in all cases and is shown in Text, Fig. 13.9. The operational amplifier controls the power fed to the load through the power transducer. Closing the feedback loop causes the sampled output to be made equal to the stable reference, without power being delivered by the reference. With high enough gain in the operational amplifier, the controlled variable can be made insensitive to wide variations in the load (load regulation) and in the power source (line regulation).

B. REFERENCES

Text, Chapter 13. Review Chapter 1, Section 7 (Power Supplies).

C. PROBLEMS

Text, Problems 13.1, 13.2, 13.4, 13.6, 13.7

EQUIPMENT

Power source: Variac (variable transformer)
12 V filament transformer
Full-wave rectifier
Filter capacitor (100 uF is sufficient for a 100-400 Ohm load.

Power amplifier: 2N3053 transistor or equivalent, with 10 K Ohm biasing resistor.

Load: 100 Ohm resistor plus 300 Ohms variable resistor (both 2 Watt).

Output sensor: 5 K Ohm potentiometer

Reference: 6 V Zener diode

Operational amplifier: 714 or equivalent.

Oscilloscope: dc-coupled, 1 mV sensitivity. A <u>differential</u> (two probe) scope is assumed below.

One of the following: Digital voltmeter, or a stable variable-voltage source (mercury battery and ten-turn potentiometer).

MEASUREMENTS

1. Connect the circuit shown in Fig. 18.1. Polarities are important. Check the polarities of the transistor, the rectifier diodes, the voltage reference, the filter capacitor, and the operational amplifier input.

Measurements of input, output, and summing-point voltages may be made with a sensitive oscilloscope. Measurements of small changes in the dc output as the load or line voltage is varied may also be made with the oscilloscope, by suppressing off most of the output voltage with a stable variable-voltage supply (or battery) so that the oscilloscope may be turned to its most sensitive dc range. Alternatively, a digital voltmeter may be used, though this is not suitable for observing the waveforms.

2. Temporarily disconnect the operational amplifier input and output and short the transistor (E to C). Turn on the Variac and observe the output voltage waveform at the load. If all is well, you should see a poorly filtered dc output. As the load is varied, the amount of ac ripple should vary considerably. (Do you recall why? See the ripple calculation, Text Section 1.7.2.) Turn off the Variac and

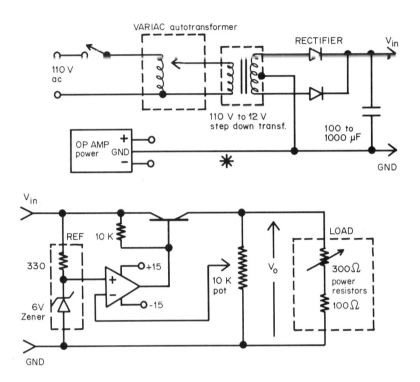

Figure 18.1 DC voltage regulator.

reconnect the transistor and operational amplifier. ALWAYS TURN OFF THE AC POWER TO THE VARIAC WHEN MAKING CIRCUIT CHANGES!

3. Turn on the operational amplifier power supply. With the oscilloscope connected at the summing point (across V_+ and V_-) and the Variac turned up two-thirds, turn on the main ac power. What happens to the summing point voltage?

4. Observe and describe the output voltage as the Variac is varied over its full range. Use an AC coupled scope to look at the AC component, and a DVM or an X-Y recorder to look at the DC component. Note, expecially, the dramatic reduction in ripple and stabilization of the DC

level when the input rises above a certain threshold. Using a differential oscilloscope, what waveform appears across the transistor? What determines the minimum output voltage at which the transistor turns on?

5. Vary the setting of the 10 K pot, which sets the fraction of output voltage fed back to the op amp. Observe and explain how the regulated output-voltage varies as the feedback fraction is varied.

6. Measure the output ripple, line regulation, and load regulation (vary the load from 100 to 400 Ohms) as described in 1 above. Lower R_L to increase the load current, to more effectively demonstrate active regulator action. A test of active regulation is that the output dc voltage varies little when R_L is varied (Load Regulation).

(Optional) Do the <u>absolute values</u> of ripple, line regulation, and load regulation make sense?

(Optional, fun for the fearless) The light regulator circuit of text Section 13.6 provides a dramatic visual demonstration of control and stability problems. The circuit of Text Figure 13.10 gives sufficient details to build it. Follow the suggestions given at the end of that section of the text.

DEBUGGING

1. Is the transistor **ON**? (test V_{ce} and V_{be} with a floating meter).

2. Is the op amp getting a sample of the output?

3. Is the feedback loop closed? (Check that the summing point really is driven to zero.)

4. Is the feedback negative? (Reference sample going to inverting input.)

EXPERIMENT 19:

ANALOG COMPUTER SOLUTIONS OF THE DAMPED HARMONIC OSCILLATOR

A. BACKGROUND

In Experiment 17, you learned about simulating mathematical operations such as differentiation and integration using operational amplifiers with various feedback connections. How you will learn how a combination of these amplifiers can simulate a physical system such as a spring and mass oscillating with damping. This simple problem was chosen because you know the answer, and, therefore, can check whether the analog solution agrees with your intuition.

This mathematical problem is called a second-order linear inhomogeneous differential equation with constant coefficients. Many interesting problems are more complicated than that, but the more sophisticated analog computers (ours is of the kindergarten variety) can handle them. For instance, if the differential equation involves the product **ty**, you generate a ramp and call it **t** and multiply it by the voltage **y**. If a combination like y^2 comes up, you need a circuit which squares a voltage. Squaring and multiplication will be covered in a Experiment 21.

Some argue about whether the analog or the digital computer is superior. That battle is not worth fighting, since it depends on the problem you want to solve. The digital computer is without competition in handling numerical data with many significant figures. But often the scientist wants to see what happens qualitatively to a system as he varies the parameters a little or adds a perturbation. Using an analog computer, one can change a few resistors to vary parameters, perhaps add an amplifier or so to add a perturbation, and the new solution is displayed immediately on an oscilloscope or x-y recorder.

There are two stages in setting up an analog solution of a differential equation. Consider the damped harmonic oscillator, Text, Eq. 14.16. An analog computer solves this problem by calling the variable **x** and its derivatives **voltages**. For instance, if you call the output of an integrator **dx/dt**, the input is necessarily **(RC)x**. The integrator time constant **t = RC** takes care of the extra time dimension and is a way of adjusting the physical parameters of the problem. When the feedback loop is closed, the analog computer begins simulating the time development of the differential equation. An appropriate combination to solve this problem is shown in Fig. 19.1 . We choose to vary the frequency $w = (k/m)^{1/2}$ by varying the ratio R_3/R_1. With R_2 and R_3 fixed for convenience, the ratio $B = R_4/R_5$ sets the amount of damping.

A second-order differential equation needs initial values of x and dx/dt to specify the solution. In the analog computer, this is done by setting the voltages at the points x and dx/dt to the desired initial values. We will do this using a relay, as shown in Fig 19.2. In the **run** mode, R_f and R_i do not affect the operation of the integrator. In the **set** mode, the capacitor is forced to charge to $(R_f/R_i)(15V)$ (Do you see why?), setting the initial value of that particular variable. This completes the first stage of setting up the problem.

The second stage of problem solving, which we will largely avoid, is called **scaling**. Suppose the input is a 10-V sine wave at w = 1 rad/sec. If RC = 0.01 sec, the output amplitude will try to be (10 V)/(wRC) = 1000 V. Since the amplifier output is limited (e.g. ±15 V), it obviously could not operate as an integrator. Scaling means extimating the ranges of all the voltages in the system and making appropriate changes in the time base (RC) or other variables to keep the solution within the voltage bounds of the computer. In this experiment, WRC ≃1, which automatically takes care of scaling problems: if x is within bounds, all its derivatives will be also.

In this experiment, you will study the response of the electronic spring and mass, with several spring constants and damping coefficients, using as initial conditions a deflection, a velocity, and a force. Those of you who have the time could look into the system's response to an oscillatory force, and also see what happens if the damping coefficient is made negative (unstable oscillations).

B. REFERENCES

Text, Chapter 14, Sections 7, 8, 9 and (optional) 10.

C. PROBLEMS

1. Text, Problem 14.5.

2. What is b/m for critical damping?

3. Sketch the expected behavior of x, dx/dt and d^2x/dt^2 for cases a, b, and c listed below:

Case	a	b	c
Initial deflection	x_o	0	0
Initial velocity	0	dx/dt_o	0
Force(t>0)	0	0	F_o

Assume that about five cycles of oscillation occur before the system damps out. In case **c**, the force term changes the system to an inhomogeneous differential equation which has as a solution both a homogeneous part and an inhomogeneous part, chosen to fit the initial conditions. Use your intuition; what will the spring do if you suddenly hang an extra weight on it?

4. (Optional) Text, Problem 14.8.

D. EQUIPMENT

Op-Amps: 4 741's, or 4 RCA CA3140's, or 4 TI TL080's
Breadboard
xy-recorder (useful, but not absolutely necessary)

> **dc-coupled scope** and **scope camera**.
> a number of **R's** and **C's**
> **2 reset relays**
> **audio oscillator**, low frequency [<1Hz] (optional).

Warning: Capacitors should be low leakage types, or else the oscillations will damp out even when no damping was intended. Polyester dielectric is best, mylar is adequate, and electrolytics are useless.

The op amps must be selected carefully to be adequate as integrators. As long as the time scale RC is 0.1 sec or less, a 741 is adequate (natural frequencies in the simulation are 10 Hz or greater). Typically choose R = 100 K Ohms, C = 1 uF in this mode. Multiple 741's in a single package are especially useful in such applications; try the 747 (two 741's) or the LM 348 (four 741's but no pins for offset adjustment). However, the speed limitations using 741's limit one to scope observations; to make xy recorder plots, it is better to stick with real time (RC = 1 sec, R = 1 M , C = 1 uf) and use a better op amp than a 741. It is strongly suggested that you try this option, using a "super 741" replacement such as the BiMOS RCA CA3140 or BiFET TL080, (see spec.'s in appendix), with 1000 times lower bias current.

Note: If you do use a 741-type, observe that when the overall system loop is closed, long term integrator offsets do not seem to ever add up to send the system out of range. Why is that?

E. MEASUREMENTS

Two suggestions: 1.) Always turn off the power before changing feedback elements; 2.) Always run an experiment with oscilloscope display before making a permanent record on the x-y recorder. This will allow you to get the sensitivity settings right to make neat graphs. Refer to Table 1 for suggested feedback component values for the following measurements:

Table 1. Suggested feedback component values

Section	1	2		3	4
R	1 M	1 M		1 M	1 M
C	1 uF	1 uF		1 uF	1 uF
R_f	50 K	50 K		50 K	50 K
R_i	500 K	500 K		500 K	500 K
R_0				10 M	10 M
R_1	1 M	1 M		1 M	1 M
R_2	1 M	1 M		1 M	1 M
R_3	1 or 10 M	1 M		1 M	1 M
R_4	10 K	10 K	100 K	10 K	10 K
R_5	1 M	1 M	200-25 K	50 K	500 K
		200 K			
		50 K			

1. Natural frequency of oscillation. Connect up the circuit of Fig. 19.1. Each integrator must have the reset/initialize circuit of Fig. 19.2 (consult wiring diagram in the lab). After checking the wiring, turn on the power and verify that the amplifiers are not saturating in either the **Run** or **Reset** mode. Remove the damping by disconnecting the appropriate jumper. Initialize the appropriate integrator to put in an initial deflection, depress the **Reset** switch, and observe the undamped oscillation at **x**. Check that the natural frequency agrees with your calculation in Prob. 1. Check the change in the natural frequency when R_3 is changed from 10 M Ohms to 1 M Ohms. What did you expect? Evaluate the ratio of spring constant to mass for the corresponding physical system.

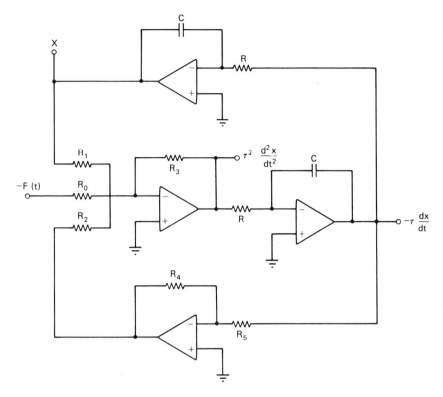

Figure 19.1 Final form for electronic analog simulation of the damped harmonic oscillator. Initial condition circuitry is not shown.

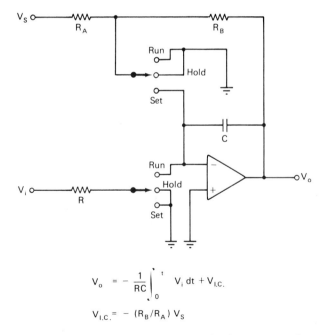

$$V_o = -\frac{1}{RC}\int_0^t V_i\, dt + V_{I.C.}$$

$$V_{I.C.} = -(R_B/R_A)V_S$$

Figure 19.2 Setting initial conditions using a three-mode integrator switch.

19-6   HARMONIC OSCILLATOR

2. Damping study. Observe and record the **x** waveform (still using an initial deflection) for various amounts of damping. Your study should include an attempt to determine roughly ($\pm 10\%$) the parameter values for critical damping, and a comparison with your calculation (Prob. 2). Evaluate the ratio of the friction coefficient (units?) to mass for the corresponding physical system.

3. Response to initial conditions. Observe and record a damped oscillation as seen in the deflection, velocity, and acceleration. Do this for an initial deflection, an initial velocity, and (with zero initial deflection and velocity) a 15 V "force" applied to the point **F(t)** in Fig. 19.1. Compare these results qualitatively with your homework sketches (Prob. 3). How big a force has been applied to the corresponding physical system?

(Optional) What sort of waveform do you expect if you give the system an initial deflection and record **x** on one axis and **dx/dt** on the other axis? Try it.

4. Forced oscillations. (Optional) Using a variable low-frequency (<1 Hz) sine wave oscillator as a "force," explore the resonant response of the system. For small damping, measure the bandwidth W, calculate $Q = W_o / W$, and compare this with the results of Prob. 4. (Optional: Use a Voltage-controlled oscillator and XY plotter whose x-axis is driven by the sweep signal, so the x-deflection will be proportional to frequency.)

5. Negative Damping: The unstable oscillator. (Optional) How would you modify the system to add a **small** amount of negative damping? Discuss your method with the instructor, and try it (with zero initial deflection and velocity) and record the result. What do you think triggers the growing oscillation?

EXPERIMENT 20: LOG AMPS

A. BACKGROUND

An amplifier is constructed and tested with logarithmic transfer function over 4 decades of input voltage. This is our first example of a nonlinear op amp circuit component, in this case a Si npn transistor in the feedback loop.

Log amp applications include measurements where the physical variables are exponentially related (example: resistivity of semiconductors as a function of temperature; chemical rate processes). With a log amp in the measurement, the plot comes out linear, so that the relevant result (energy gap or activation energy, in these two cases) may be simply and accurately obtained. Sometimes one uses a log to allow a wide dynamic range to be spanned on a single output device, for example in a vacuum gauge where one wishes to roughly follow the progress of a pumpdown over many decades of pressure.

The principal problem in designing a log amp is drift, due mostly to the input bias current I_b of the op amp. The output voltage error $V_o = I_b R_f$, where R_f is the (nonlinear) resistance of the feedback transistor. Since R_f may reach 10^8 Ohms, this error can be quite significant. A second source of error is the temperature sensitivity of the transistor, which alters the scale factor kT/q.

An-FET input op amp is used to minimize the first problem. In more exacting applications, one would in addition replace the transistor by a circuit utilizing dual log transistors to cancel most of the thermal drift.

B. REFERENCES
   Text, Chapter 15, Section 3.

C. PROBLEMS
   Derive the transfer function of the circuit in Fig. 20.1. The result will be of the form

$$V_O = a + b \ln V_{IN}$$

Referring to the figure, what variables determine **a** and **b**? What device variables determine **a** and **b**? For example, what can one manipulate to make the circuit cover the desired dynamic range (2 decades - or 4?) of input, and a convenient range of output?

D. EQUIPMENT
   Log-amp circuit (Fig. 20.1).
   A **1.5 V** and a **90 V battery** are used to generate input signals. Two 10-turn pots are needed, with resistance value about 10 K Ohms.
   A **wide range voltmeter** is needed for V_{IN}; a multimeter (VOM) is fine. A high resolution voltmeter is needed for V_{OUT}. A 3-1/2 digit DVM is desirable, since the FULL range of V_{OUT} is only about 0.5 V!
   A **scope** or **XY recorder** is useful to quickly record rough behavior. However, accurate values of V_{OUT} must be written down for plotting on semi-log paper.

<u>CIRCUIT DETAILS AND SUBTLETIES</u> (Refer to Fig. 20.2).
   The op amp: AD 506 (Analog Devices), an FET input op amp. Its main features are very low offset current (~10 pA); fairly low offset voltage (~1 mV); low voltage drift (~10 uV/°C). A 741 will just not work in this application! Note that there are 7 wrong ways and only one right way to insert the leads on an AD506. Be careful (> $10 each). Alternative choice: CA 3140, or AD 547 BiFET or other BiFET type are lower cost solutions.

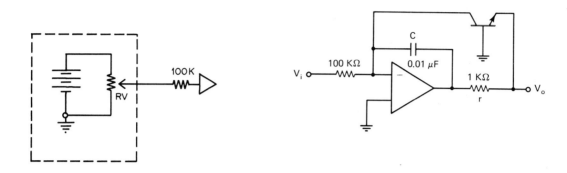

Figure 20.1 Practical-stabilized log amp circuit.
Important: Offset nulling not shown. Consult Fig. 20.2.

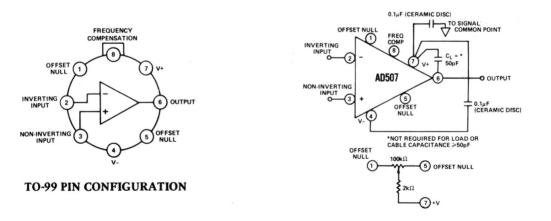

Figure 20.2 Pin Diagrams

LOG AMPS   20-3

The bias current is roughly balanced with R_b (on the + input) of about 100 K Ohms. Vary if needed. A capacitor C = 0.2 uF across the transistor helps to stabilize the circuit against high frequency oscillations. But if you use the capacitor, a large resistor (10 M to 100 M) is also needed in parallel to keep the output drift down.

For offset voltage balancing, it is **essential** to use a multi-turn pot, 10 turns or more. The value is not critical; 10 K is shown, but 20 K will also work. The input voltage has to span a wide (~ 4 orders of magnitude) dynamic range. Therefore, R_v must be a multi-turn pot (10 or more turns).

It also helps to use several values of V_{IN}. Use a 1.5V battery to access the 10^{-2} to 1V range. Then use a total of 90 V to go to the high input range. (As long as the battery just establishes an input **current**, one doesn't need to worry about it being so large. But BEWARE: if 90 V is applied **directly** across its inputs, the op amp will be destroyed.

Concerning our choice of transistor: the SK3021 is nothing magic. The transistor should be Silicon, and npn for this choice of input polarity. A power transistor is used merely to create an I-V curve with fairly large I for a given V. That reduces the effective value of Z_f, and makes less demands upon the op-amp's stability and offsets.

The input polarity matters. For the circuit shown; the input must be positive. What if the input voltage is reversed? The collector-base junction will get **forward** biased. Maybe it won't hurt anything, but it won't be a good log amp.

E. MEASUREMENTS

1. First, wire up the op amp in a gain of 100 inverting amplifier. With $V_{IN} = 0$ (be sure; short to ground), balance the offset voltage to zero. For an AD 506, this involves the offset pot (pins 1 and 8).

2. Connect the circuit shown in Fig. 20.1. Monitor V_{IN} with a multimeter. Monitor V_{OUT} with a digital voltmeter. If a DVM is unavailable, one will need a variable and

calibrated zero suppress circuit of some kind. An initial warmup (5 min) will likely be needed before transistor and op amp stabilize. With such a high Z_f, the circuit will be sensitive to various capacitative coupling (including yourself). Balance the offset such that it stays balanced when you are <u>not</u> touching the circuit.

3. Vary V_{IN} over a wide range to get a rough idea if the circuit is working. V_{OUT} does not change much (it goes as the log!). In our version, a 4 decade change of V_{IN} changes V_{OUT} only from 0.3 to 0.6 V. A quick permanent record of the log response may be made using an xy recorder or an xy oscilloscope. Vary V_{IN} from about 90 V down to about 10^{-3} V with about four points per decade (a 1,2,5,10 sequence comes out evenly when plotted on a log scale). Plot the results with V_o (range 0.30 to 0.60 V) on the x-axis and log I_{in} (range 10^{-8} to 10^{-3} A) on the y-axis, where $I_{in} = V_{in}/100$ K. We suggest that you plot I_{in} rather than V_{in} in order to facilitate understanding of any deviations from perfect log behavior at the low end.

APPLICATIONS (Optional)

What can you use a log amp for? Anything where the input covers an extremely wide dynamic range, and you want to display it on a single meter or chart.

Examples you might try:

a. Pressure measurement in a vacuum system. The log of the output of an ionization gauge gives you a way to monitor the entire pumpdown process.

b. Semiconductors have resistivity which varies exponentially with temperature. A log amp makes it easy to plot on an xy recorder.

c. The even-tempered scale is generated on an analog electronic music synthesizer using a linear voltage divider followed by a log amp.

EXPERIMENT 21: MULTIPLIERS

A. BACKGROUND

The multiplier is easy to use but surprisingly powerful in its possibilities. The emphasis in this lab is therefore not at all on circuit complexities but rather on imaginative applications of this nonlinear element. Much of analog electronic music develops from the multiplier, and you will find this lab more enjoyable if an audio amplifier and speaker (and tape recorder) are available. The analysis of some of the observations is subtle, and will be greatly aided by some background in Fourier spectral analysis. You will surely need to be familiar with the trigonometric identities for the products of two sine waves, such as:

$$\sin w_1 t \cdot \sin w_2 t = \sin(w_1 + w_2)t + \sin(w_1 - w_2)t$$

since the sum and difference frequencies will be quite prominent in your explanations. You will also need to look up the sine wave harmonic Fourier components of some simple waveforms: triangle and square wave.

B. REFERENCES

1. Text, Chapter 15, Sections 5-8.
2. Analog Devices Specification Sheets for the multiplier used (e.g., AD532 or AD533).
3. Daniel H. Sheingold, ed.: <u>Nonlinear Circuits Handbook</u>.

C. PROBLEMS

1. What frequency spectrum do you expect at the output of a multiplier in the following cases? What will it look like as a function of time. The **X** and **Y** inputs are both sine waves, with frequencies specified in Hertz.

X	Y	Out
1000	1100	?
1000	100	?
1000	100	?

Hint: see text, problem 15.3

2. Text, Problem 15.4, 15.5.

D. EQUIPMENT

1. **IC Transconductance Multiplier.** Analog Devices AD 533 (or AD 532). The differences are mainly in ease of use. The fewer external trimpots required, the more expensive the multiplier. The AD 532 has differential input capability and is internally trimmed. It is relatively expensive (~$25). The AD 533 is an inexpensive (~$5) compromise. It has 1% linearity and is self-contained (no extra amplifier needed) but does require three trimpots top adjust offset, feedthrough and symmetry. We <u>strongly</u> recommend that you use the inexpensive 533. Multipliers are fairly easy for the inexperienced user to burn out.
2. **Audio oscillator.**
3. **Function generator.** Voltage controlled if possible.
4. **Pulse generator.** Width and repetition rate in the audio range.
5. **Oscilloscope and camera.**
6. (A Hi-Fi amplifier and speakers are also nice.)

CIRCUIT PREPARATION

Wire up the IC multiplier as shown in Fig. 21.1. Refer to the AD specification. Proceed approximately with the trim procedures. Don't worry much about the full scale gain adjustment. You will also find that the three pots can easily be adjusted to optimize the DC offset, feedthrough,

and symmetry. Use a test input consisting of:
   X: sine wave, 1000 HZ
   Y: pulse or square wave, 100 HZ.
Check your circuit carefully! Verify that it works as a multiplier using several combinations of DC voltages.

E. MEASUREMENTS

Take pictures of waveforms you find to be typical of multiplier signal manipulation. An audio amplifier and speaker can add interest; can you figure out why the signals sound as they do? Do the sounds and scope displays of a given signal both make sense?

An especially interesting array of outputs can be generated when one of the inputs is a <u>swept</u> voltage controlled oscillator. A little imagination can result in a variety of very complex electronic music. You may wish to tape some of your best sounds, along with explanatory comment.

Scope triggering can be difficult with two incommensurate frequencies. Try external triggering from the pulse or one of the sine waves. A slight adjustment of the second frequency can then stabilize a pattern.

1. Simple Pulse and Sine Wave Combinations

Adjust the pulser (V_x) in the range of 100 HZ repetition rate and pulse width about 3-5 msec. (Important: To see the most interesting results, the pulser voltage must go to zero between pulses, rather than being symmetric with respect to ground.)

MAKE SURE BEFORE CONNECTING ANY INPUT THAT ITS VALUE DOES **NOT EXCEED 10 V** OR ELSE THE MULTIPLIER MAY BURN OUT. Also, recognize that the product XY will necessarily be limited to the supply voltage, so larger input products will result in output distortion or clipping.

With circuit gain adjusted so that $V_o = (V_x \cdot V_y) / 10$, however, V_x and V_y may reach 10 V without distortion. That convenient scale factor may be set by choosing the gain resistor (Fig. 21.1), about 6 K.

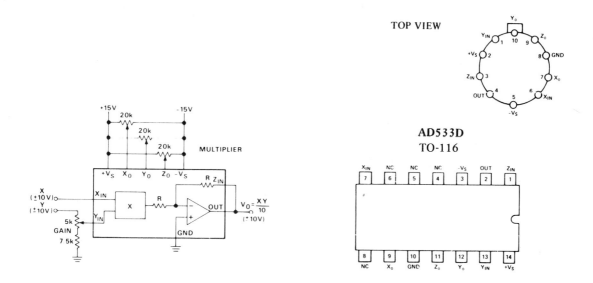

Figure 21.1 Multiplier connections for the AD533.

Adjust the oscillator (V_Y) to about 1000 Hz. Observe the output signal. Adjust the offset pot to eliminate the DC output. Adjust the feedthrough pot so that no sine wave (or a minimum amplitude) is seen during the off-time of the pulser. Adjust the third pot until the sine wave part of the output is symmetric with respect to the ground.

This output is called a tone burst, and is useful in checking transient response of audio components.

(Listening) Lower the pulse width. How does the sound of the 1000 Hz "tone" change? This test is easier if the pulse rate is slowed to 1 HZ or so. Find a pulse width below which you can no longer distinguish the pitch of the sound. Explain!

Now reverse the frequency ratio so the sine wave is about ten times slower than the pulse rate. Note the "chopped" sine wave output.

21-4   MULTIPLIERS

2. Simple Sine Wave Combinations

Using two oscillators, observe the possibilities for combining tones.

1. Put the same oscillator into both X and Y. What <u>frequency</u> comes out? Why?

2. Try $f_2 \simeq 10\, f_1$. Note the "modulated" output. In amplitude modulation, this product is called <u>suppressed carrier</u>. Why? What is the frequency spectrum?

3. Now try $f_2 \simeq 1.1\, f_1$. Note the appearance of a low frequency component. Where does it come from? If you make f_2 approach f_1, what happens to this component? The output of a multiplier with two equal frequencies is the basis of coherent detection (See Exp. 25, Lock-in Amplifier; Exp. 27, Phase-locked Loop).

3. Voltage Controlled Function Generator Plus Pulse Or Sine Wave

Repeat 1 and 2, but allow the frequency of one sine wave to be swept as a function of time. An extraordinary range of patterns can be generated on the scope. If you have an audio output available to listen to here, one's imagination can lead to "electronic music."

4. Further Possibilities (OPTIONAL)

1. Let the function generator output be a triangle or square wave. Observe the combination tones of the harmonics in the triangle or square waves and a second input (sine or pulse).

2. Make a divider. Use DC signals for inputs, and see how linear the divider is. Look for problems at the high output end.

3. Check the basic multiplier linearity. DC signals are most accurate, but a rough display can be obtained with an XY oscilloscope. Using dc signals, a quick but accurate record can be obtained using an XY recorder.

EXPERIMENT 22: ACTIVE FILTERS

A. BACKGROUND

You will construct several kinds of active filters useful in signal processing. A **low-pass** filter is useful in removing high frequency noise. One can select a circuit for various rolloff rates (first order : 6 dB/octave; second order: 12 dB/octave), and can adjust the shape of the transfer function to optimize the fidelity within the signal range of interest. Several versions of **band-pass** filter are available. Putting a twin T in the feedback loop gives a narrow bandpass. A more flexible and clever bandpass filter uses two op amps in the feedback loop of a third, simulating an inductor. The result, the biquad filter, simulates a parallel RLC tuned circuit.

B. REFERENCES
Text, Chapter 16.

C. PROBLEMS

1. Analyze the 2nd order low-pass filter (Fig. 22.1). Derive the transfer function, Text, Eq. 16.4. What is the function of the variable b?

2. Analyze the biquad band-pass filter, text Section 16.4.4. Obtain expressions for the resonant frequency, the Q, and the gain. What is the procedure for setting circuit values to obtain a specific resonant frequency, Q, and gain?

D. EQUIPMENT

For audio work, 741, 747, or LM324 general purpose op amps will be adequate. In a high frequency application, one would switch to a wider bandwidth uncompensated op amp.

A frequency counter and digital multimeter with ac voltage capability can be used together instead of a scope for obtaining numerical values of transfer functions quickly.

(Optional) In mapping transfer function, you may choose to avoid tedious point-wise data taking by using a swept voltage-controlled oscillator to scan the frequency range. A scope photo is an adequate record of the transfer function. As an alternative, use an xy recorder, with the y-axis driven by a rectified and filtered version of the filter output. (See text Fig 15.3.)

Select cutoff or center frequencies of your filters below a few kHz to avoid complications due to the op amp bandwidth.

E. MEASUREMENTS

1. Active Low-Pass Filter. Construct and test the first order and second order filters (Figs. 22.1 and 22.2). Save time by having both set up at the same time on the same breadboard, to aid in comparison.

Testing: Include both **sine wave** and **step response**.

Sine wave: Map the transfer function (ROUGHLY)

a. Verify that the cutoff frequency is given by the appropriate $(RC)^{-1}$, and that it can be varied at will by varying R, <u>without</u> changing other filter characteristics.

b. Check the rate of rolloff for both filters; 6 dB/octave or 12. The cutoff frequency is conventionally measured as the point where the voltage output falls 3 dB (factor of 0.707 in voltage or a factor of two in power) below its low frequency value. The asymptotic slope rolloff rate is not generally achieved until well above this value, e.g., until the output is down by 10-20 dB.

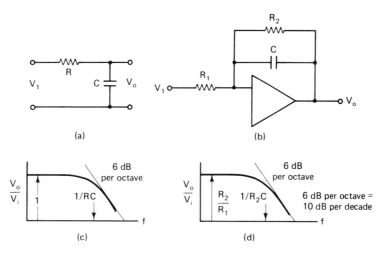

Figure 22.1 First-order low-pass filter.

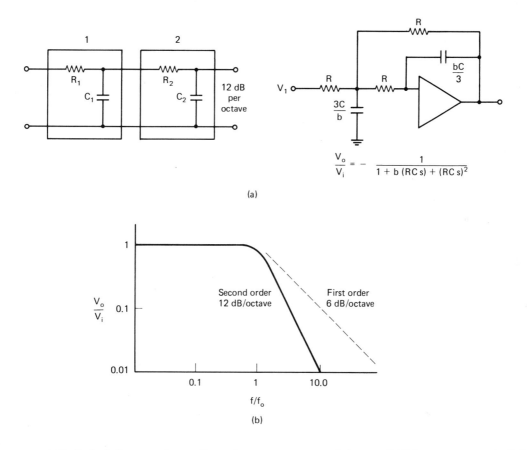

Figure 22.2 (a) Second-order low-pass active filter on right, with improved performance over the passive version shown on the left; (b) Transfer function of the active second order low pass.

ACTIVE FILTERS   22-3

c. (Optional) Explore the output when the input signal includes higher frequency "noise." Examples: 100 Hz sinewave plus amplified speech or background noise; sinewave "signal" plus another high frequency sinewave "noise." Measure on the oscilloscope the improvement in signal-to-noise amplitude brought about by the filter.

Step Response: Use a square wave or a pulse generator. Explore the step response of both filters. Pay particular attention to comparing the "rise time," the time to change by $(1 - 1/e) \simeq 2/3$ of the step height. In the case of the second-order filter, the extra variable b controls the existence of overshoot or undershoot. Explore this, and (going back to sine waves), see what connection there is between the sine wave transfer function, and the step response, when the step response is overdamped or underdamped.

2. Active Bandpass. Select **a** or **b** (**b** is most fun).
a. Twin-T. Construct a twin T notch (Fig. 22.3). Use components as closely matched as possible (1% resistors, for example), or allow for some adjustment to make the notch as deep as possible. Select a center frequency f_o low enough so op amp rolloff does not interfere. Measure the depth of the notch. Why is it not perfect? Measure the transfer impedance at f_o. (How? Recall from the text that the ratio of Voltage input on the left to current output at the right.)

Put the twin T in the feedback loop. Watch out for oscillations. Use an extra R in parallel with the T if necessary to quench oscillations.

Map the transfer function for sinewaves. Now observe the step response.

b. Biquad. Construct a biquad bandpass filter (Fig. 22.4). Select a center frequency low enough so op amp rolloff does not interfere. See how one can vary f_o, the bandwidth and corresponding Q, and gain G with single components. Do the values agree with what you expect? Map the transfer function and observe the step response for several values of Q.

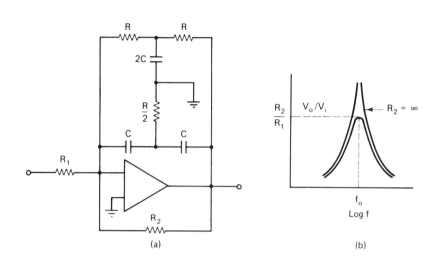

Figure 22.3 (a) Twin-T filter circuit; (b) The twin-T's transfer function.

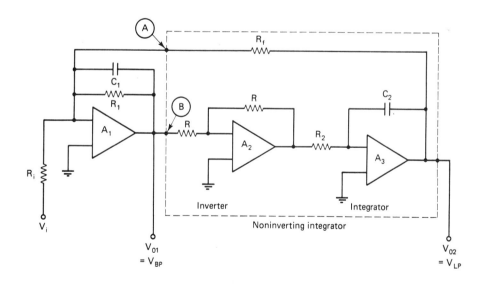

Figure 22.4 Biquad bandpass active filter. The dashed box acts as an active inductor viewed between terminals A and B.

ACTIVE FILTERS   22-5

(Optional) Pass a square wave through the bandpass. Vary the frequency and observe peaks in the output amplitude as the various harmonics pass through the resonance. (An audio output is useful here, as the experiment generates an interesting electronic music sound.) If the square wave frequency is slowly swept and the output displayed on an oscilloscope or recorder the resulting array of peaks is a measure of the relative amplitude of the various harmonics. Check the numerical values. This is the principle of analog spectrum analyzers.

NOTE ON BANDPASS MEASUREMENTS

A frequency counter is of great use in quickly measuring Q, and is <u>essential</u> when Q is high (>20).

To simplify varying R and Q in the Biquad, use pots, preferably multi-turn.

EXPERIMENT 23: OSCILLATORS AND INSTABILITY

A. BACKGROUND

Since one can buy an oscillator cheaply, why a lab on this subject? Analog circuits can oscillate unexpectedly. The rationale of this experiment is thus not only to become acquainted with some standard oscillator circuits but, more importantly, to understand the conditions under which an oscillation can build up. Two classes of systems will be studied.

1. Positive feedback oscillators: the feedback comes back in the correct phase to build up and sustain the oscillation. Pay attention to why one particular frequency is favored and others suppressed.

2. Tuned circuit oscillators: the feedback network has a characteristic resonant frequency.

B. REFERENCES
Text, Chapter 17, first half.

C. PROBLEMS
Text, Chapter 17, Probs. 17.1, 17.3, and 17.4.

D. EQUIPMENT
Op amp. 741 or equivalent.
Audio oscillator.
Oscilloscope and **camera.**
Passive components as shown on circuit diagrams.

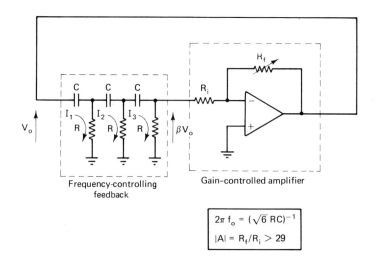

Figure 23.1 Circuit diagram for the basic RC phase shift oscillator.

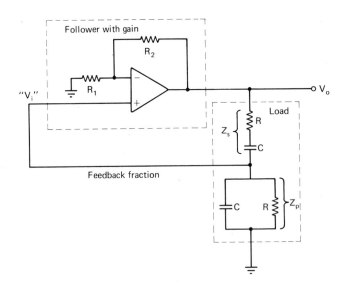

Figure 23.2 Wien bridge oscillator circuit.

23-2   OSCILLATORS AND INSTABILITY

E. MEASUREMENTS

1. Select **one** of the following:
   Phase shift oscillator (Fig. 23.1(a));
   Wien bridge oscillator (Fig. 23.2).

   Select component values and with a frequency in the neighborhood of 1 kHz. Vary the amplifier gain (R_f/R_i in Fig. 23.1; R_2/R_1 in Fig. 23.2) until the oscillations build up and become stable. Now open the feedback loop and measure the gain of the amplifier portion alone with a commercial oscillator (at that same frequency) as an input. With the same oscillator as input, measure the attenuation of the impedance network alone at the critical frequency. Use these results to show how well the oscillator criterion (B a >> 1) is satisfied.

   Explore waveforms at other places in the circuit. In particular, look at the output of the feedback network. It will be typically quite distorted from a sine wave. How is it possible that the oscillator output from that distorted input is such a clean sine wave?

   If your circuit includes a diode clamp limiter [Text, Fig. 17.9 (a)], observe the signal on either end of the diodes. How is it possible that this very distorted waveform serves to limit the amplitude of the sine wave without appreciable distortion? Remove the diode and replace the 1 K resistor with a 10 K pot. Vary the feedback fraction, and observe how narrow a range results in oscillation without clipping.

   (Optional) Explore the use of a current-sensitive resistor (very low wattage tungsten light bulb) in the Wien bridge circuit to stabilize the amplitude with minimum distortion. How stable is the amplitude? Try suddenly changing a component which sets the natural frequency. How pure a sine wave results (compared to a diode clamp)?

2. Select **one** of the following:
   Twin-T oscillator (Fig. 23.3);
   LC tank circuit oscillator (Fig. 23.4);
   Crystal Oscillator (Fig. 23.5).

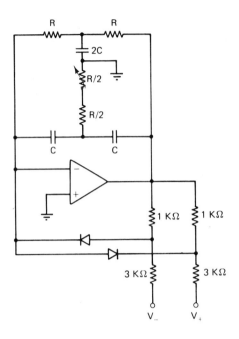

Figure 23.3 Working circuit of twin-T oscillator, with diode limiting of amplitude.

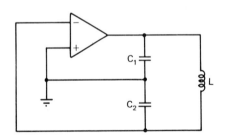

Figure 23.4 LC resonant circuit or Colpitts oscillator.

Explore the frequency response of the feedback network by itself. This is easy for the **twin tee**. For the **LC tank circuit** or **quartz crystal** options, study the parallel-resonant behavior, measuring the voltage as a function of frequency when the circuit is fed by an (approximate) current source: oscillator plus 1 to 10 MOhm resistor in series (consult your instructor if necessary). Note

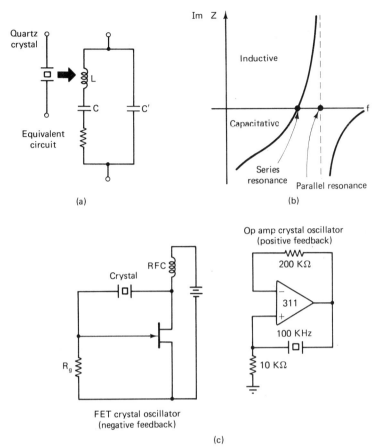

Figure 23.5 Quartz crystal oscillator or clock.

carefully what the natural frequency is, and any phase shifts near that frequency. Use a frequency counter to pin the frequency down accurately. Now connect the network into the feedback loop, and vary the gain until oscillation is observed. What is the frequency? Is it exactly equal to the natural frequency of the network? Given the phase shift of the op amp in the connection used, does the phase shift of the feedback network add up to a net positive (<u>regenerative</u>) feedback?

Observe the voltage waveforms at both the input and output sides of the feedback network, and comment on the waveshape observed.

In both parts 1 and 2, compare the frequency you observe with what was expected for the component values used.

EXPERIMENT 24: FUNCTION GENERATORS

A. BACKGROUND

As in Exp. 23, the purpose of this experiment is not just to build a function generator, since commercial IC's exist for that purpose at modest cost. There are some new circuit principles to be learned.

1. The function generator combines linear op amps with a two-state switching circuit, the comparator. Consider the right hand op amp in Fig. 24.1. It looks like a conventional amplifier with unity gain. But look where the feedback connection is made: to the noninverting input. This positive feedback adds hysteresis or memory to the circuit, whose basic function is to be driven to full positive or full negative output, depending on the sign of the voltage across its input terminals.

2. A second non-linear function is added in the voltage-controlled function generator (Fig. 24.2), with the multiplier, which sets the amplitude of the integrator's ramp speed and hence the frequency of the circuit.

3. A commercial IC voltage-controlled oscillator offers not only a ready source of useful waveforms, but also a means to convert a physical quantity such as temperature, pressure, or position, into a high-precision measurement.

B. REFERENCES
Text, Chapter 17, second half.

C. PROBLEMS
Text, Probs. 17.7, 17.8, and 17.9.

D. EQUIPMENT

Oscilloscope.

Frequency counter.

Op amps (2). For audio frequencies, 741's will do for both. For lower frequencies, the integrator portion needs a lower bias current. Try a BiFET (TL080, for example). For higher frequencies, a high-frequency comparator op amp (LM311, for example) is desirable.

Multiplier: AD533 or equivalent.

IC voltage-controlled function generator: 555 or 8038.

Transducer (optional) such as a thermistor.

E. MEASUREMENTS

1. Simple Function Generator

Wire up the circuit shown in Fig. 24.1. Select R and C to give f = 1 kHz. First wire up the integrator and comparator separately, so that their operation can be tested without the complication of cross-coupling. You will need to balance to zero the offset voltage of the integrator op amp, with input grounded. It is simpler to make a non-ideal integrator with an additional 1 Megohm resistor across the feedback loop. If you cannot balance the offset, try balancing it first as a gain-of-100 amplifier, or get a better op amp. Now explore the operation of the comparator. When the power is turned on, what happens to the output? Can you change the state of the comparator? Apply input voltages of $\pm V_{CC}$.

Now interconnect the comparator and integrator. If working properly, the circuit should oscillate. (The Zener diode limiter is essential to prevent latch-up.) Measure the period, and compare it to what you expect for your values of R and C.

Vary R and C. Does f scale as expected? Measure the rise time of the square wave. Does it agree with what you expect for the op amp used as a comparator?

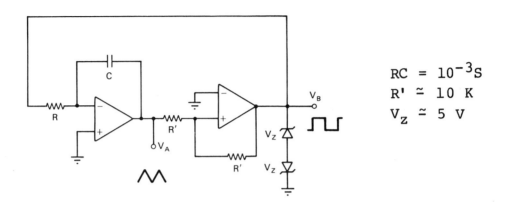

Figure 24.1 Basic function generator circuit.

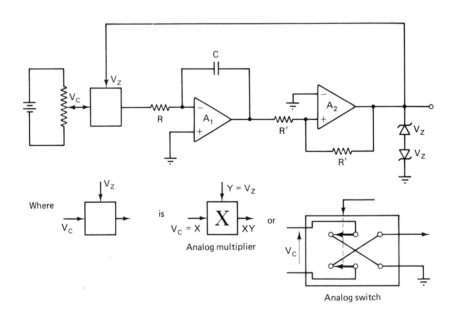

Figure 24.2 Voltage controlled function generator.

FUNCTION GENERATORS   24-3

2. Voltage Controlled Oscillator

Wire up a circuit such as shown in Fig. 24.2. Use a multiplier to provide the control function, and make sure that V_Z is within the multiplier range (typically $\pm$ 10 V). Vary the control voltage, and plot the frequency as a function of voltage. Do the values and range make sense? Check and interpret the waveforms at the output of each op amp.

3. (Optional) Integrated Circuit VCO's

Wire up and test one of the IC VCO's (556: Text, Fig. 17.18, or 8038: Text, Fig. 17.19). Verify that the operation varies with external components and with control voltage as expected. Refer to detailed spec. sheets (Appendix) as needed. In the case of 8038, explore the quality of the sine wave. Can you think of a way to see the breakpoints of the diode function generator? [Hint: differentiate.]

4. (Optional) Measurement Instrumentation

Make use of the fact that these circuits allow you to convert any physical quantity for which a resistive, capacitive, or voltage transducer exists, into a frequency which can be measured to high precision.

The transducer can take the place of the integrator R or C, or of the VCO voltage input. Example: temperature measurement. Use a temperature-dependent resistor: a thermistor or even an ordinary carbon resistor, connect the function generator output to a frequency counter. Observe and record the frequency shifts at calibration points: ice-water and boiling water. It is best to put the transducer in a test tube to avoid errors due to the conductivity of the water. If the sensitivity of your circuit is sufficient, you can now monitor very small temperature changes with great precision.

EXPERIMENT 25:

LOW LEVEL SIGNALS AND HIGH PERFORMANCE OP-AMPS

A. BACKGROUND

The purpose of the experiment is to gain familiarity with an example of a high performance op amp, and with the kinds of problems (and their solutions) which one encounters when working at the microvolt level. Interesting low level sources include:

Johnson noise from a resistor (white noise);

Flicker noise from a back-biased transistor (pink noise);

Thermoelectric voltages from two dissimilar metals in a temperature gradient;

Bioelectric signals from muscle, heart, or brain activity. An example is shown in the text, Fig. 18.1.

Select one of these three classes of sources. For the source you select, select the one of the circuits in Fig. 25.1 most suitable for that source, and in addition the instrumentation amplifier circuit of Fig. 25.1(d). Pay special attention to differences you observe between the behavior of the two amplifier configurations (dc drift, common mode problems, or source loading). It will be essential to follow the circuit precautions shown in Fig. 25.2, for the op amp which forms the lowest level stage of amplification.

B. REFERENCES
Text Chapter 18, Sections 1 and 2.

C. PROBLEMS
Text, probs. 18.1, 18.2, and 18.3.

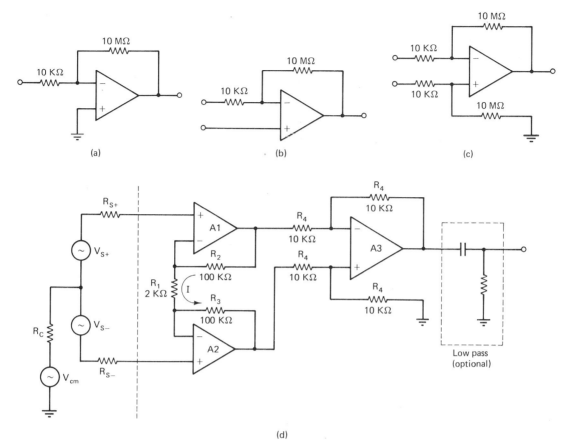

Figure 25.1 Possible ways to measure a small signal. (a) Inverting amplifier; (b) Follower with gain; (c) Subtractor; (d) The instrumentation amplifier circuit; A_1, A_2 are low noise, low offset, low bias current premium op amps.

D. EQUIPMENT

Noise sources: see Fig. 25.3.

Electrode paste: salt/glycerine.

Thermocouple: either a standard A-B pair such as iron/constantan or an Fe wire plus Cu wire.

Strip chart recorder or **xy recorder** with time base.

Thermoflasks, two: one for ice water and one for boiling water.

A willing human subject **[note WARNING in Section E.3]**.

Op amps: Standard 741's except for the low-level (first stage) where low-offset, low-drift types are essential. Guaranteed good results: try Analog Devices AD504.

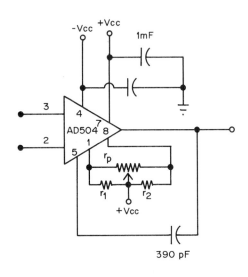

Typical Values

$r_1, r_2 \simeq 10K$
  (coarse nulled)

r_p = 100 K trim pot
  (fine nulled)

Figure 25.2 Some **extra** subtleties in using low level op amps. The AD 504 is shown as an example. These include: power supply bypass capacitors connected physically near the op amp; offset voltage nulling (pins 1 and 8); output frequency compensation capacator (pin 5).

Extra care is needed in circuit construction in order to reduce both dc offset and ac noise to the levels which such an op amp is capable of. This is shown in Fig. 25.2. To avoid noise (particularly at line frequency) brought in on the power leads, bypass capacitors are used on the circuit board as close as possible to the low level op amp. Offset nulling to the microvolt level requires very fine adjustment. This is most simply done by first nulling the dc offset in the usual way with a 10 K pot, then measuring the resistance values on either side of the center terminal. Resistors of these values are then wired in to provide a permanent coarse null. Fine nulling is accomplished by a much larger pot (100 K Ohm) in parallel.

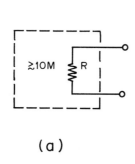
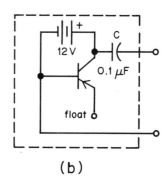

Figure 25.3 Noise sources. (a) Johnson or white noise; (b) Pink noise from back-biased transistor.

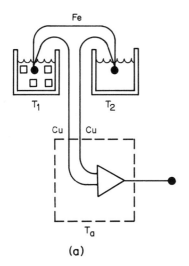
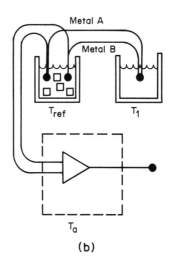

Figure 25.4 Thermocouple connections. (a) Basic idea; (b) Reference junction technique.

25-4   HIGH PERFORMANCE OP-AMPS

E. MEASUREMENTS

1. Noise experiment. Connect a carbon resistor to the input terminals of a single ended follower-with-gain or the instrumentation amp. How can you tell if you are seeing Johnson noise? (Vary R and/or T.) Look at the output when the noise source resistance varies from 1 Ohm to 100 Megohms. Use a scope triggered on "automatic" or "free running" to see the broadband noise. It will be essential to devise a shorting arrangement at the input to get an estimate of amplifier noise. But the shorting arrangement must not act as an antenna!

It is interesting to compare the spectrum of Johnson noise with the pink noise from a back-biased transistor (collector-base junction). This can be seen roughly on an oscilloscope. It can also be heard, by connecting the output to an audio amp and speaker.

2. Thermoelectric Experiment. Thermoelectric voltages are of the order of 10 uV per degree C. They provide a reliable way to measure small temperature differences. They also often get in the way of low-level dc measurements, when a thermocouple is set up accidentally! A thermocouple is set up whenever two junctions between dissimilar metals are at different temperatures. The simplest form uses two copper wires run out from the amplifier, with a wire of another metal completing the circuit, as shown in Fig.25.4. The second wire may be anything which is available besides Cu. If wires of alloys conventionally used for thermocouples are available, then readily available conversion tables may be used to convert the measured voltage to a temperature difference. In the more conventional thermocouple [Fig. 25.4 (b)]. The two metals A and B run as a pair between the two temperatures, with ordinary Cu wires to the amplifier. One of the junctions (reference junction) is immersed in ice water. It can be shown that the copper lead wires will introduce no additional thermoelectric voltages, provided both reference junctions are at the same temperature, and that the amplifier terminals temperature is kept uniform and constant.

3. Bioelectric signals. Start with the heartbeat, using wires taped lightly to the backs of the hands, with a salt solution providing good contact. A third ground electrode is necessary to reference the body to the op amp ground. This can be any convenient point such as a clothes pin to the ear.

WARNING: **EXTREMELY LOW CURRENTS CAN BE LETHAL WHEN THE CURRENT FLOW PATH INCLUDES THE HEART. BE EXTREMELY CAREFUL TO CHECK THE CIRCUIT WIRING AND AVOID THE POSSIBILITY OF POWER SUPPLY VOLTAGES APPEARING AT THE INPUTS DUE TO A WIRING ERROR OR TO LEAKAGE CURRENTS. NEVER DO A BIOELECTRIC MEASUREMENT WHILE ALONE.**

The heartbeat has a characteristic "signature" which is useful in diagnosis of heart problems. Can you see it? A strip chart recorder is necessary for this experiment. Heart signals are in the millivolt range. They can be obscured by much larger signals from arm muscles, which show up (at much higher frequencies) when the muscles are tensed.

Brain signals are several orders of magnitude smaller than heart signals. To see them, you will need to pay careful attention to making a noise-free contact. Small silver plate electrodes and an electrode paste (glycerin and salt) are useful. The two differential electrodes should be taped at the front and back of the head at one side. The ground electrode may be at the ear or neck.

Notice the characteristic changes in the spectrum when the eyes are open or closed. (The large spikes are muscle signals due to the eyes.) With eyes closed, try various modes of thought or relaxation, and see if you can observe changes in the characteristic frequencies.

EXPERIMENT 26: LOCK-IN AMPLIFIER

A. BACKGROUND

The lock-in amplifier is one of the nost powerful instrumentation circuits which you will encounter due to its ability to recover small signals from much larger noise. Either of the two circuits below will provide an adequate introduction to lock-in operation. Circuit II can be incorporated into an instrument capable of recovering microvolt level signals.

B. REFERENCES

Text, Chapter 18, Section 3.

C. PROBLEMS

Text, Problems 18.4, 18.5, 18.6, and 18.7.

D. EQUIPMENT

Lock-in Amplifier circuit [Either version A (Fig. 26.1) or version B (Fig. 26.2)].
Oscilloscope
Two audio oscillators
Frequency counter
Digital multimeter

E. MEASUREMENTS: Select I or II. Circuit III is an optional project extension.

1. Lock-in Circuit I: Diode Switching. This circuit, though far from being a state-of-the-art lock-in, can be constructed from readily available components, and, being

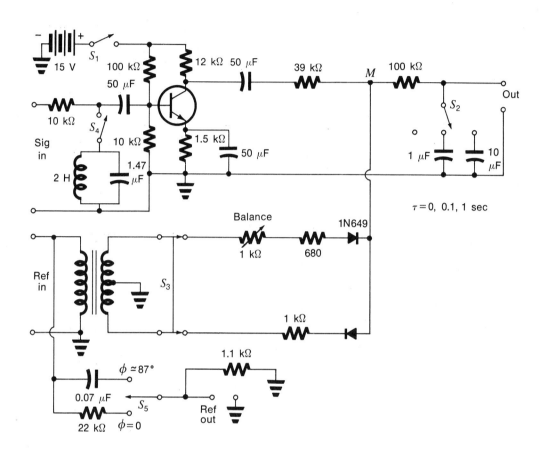

95-cps lock-in amplifier. S_1:power switch; S_2:time-constant selector switch; S_3:reference input switch; S_4:signal tuning switch; S_5:phase selector switch.

Figure 26.1 Diode Switching Lock-in.

26-2   LOCK-IN AMPLIFIER

Table 1

Part	1	2	3	4
Switches:				
Sig Tuning	on		on/off	on
Ref In	off	on	on	on
Ref Out Phase				0,90
Time Constant	0	0	0,0.1,1	0,0.1,1
Connections:				
Sig In	Osc 2	Open	Osc 2	Ref Out
Ref In	Osc 1	Osc 1	Osc 1	Osc 1
Ref Out	Open	Open	Open	To Sig In
Output	Scope	Scope	Scope	Scope

discrete, allows one to look inside and to see non-ideal effects (e.g., the reference balance condition). The circuit, shown in Fig. 26.1, is functionally similar to the diode switch [Text, Fig. 18.4(b)]. Additional features include a simple tuned preamp, a variable RC time constant, a variable reference balance pot, and passive phase shifter to provide either a 0-degree or (nearly) 90-degree phase shift of the reference signal. Perform the measurements described below with switch settings listed in Table 1. In addition to the lock-in, you will need two audio oscillators and an oscilloscope.

Note: Oscillator 1, the reference, is always set at f_o = 95 cps, 10-V peak to peak. Oscillator 2 is set at a variable frequency, and amplitude about 1 V or less. Make sure that the oscilloscope connected at the output is dc-coupled. When the time constant is zero, this connection also looks at the mixer waveform (point M).

a. Tuned-amplifier bandwidth: Sweep Osc 2 through the resonant frequency and observe the ac voltage at the output. Measure the bandwidth over which the output falls to $(1/2)^{1/2}$ of the peak value. This will be useful in comparison with the lock-in bandwidth.

b. Reference balancing: Vary the balance control until a zero output is observed. Sketch and explain the ac waveform observed as the control is adjusted on either side of balance.

c. Incoherent lock-in action: With the signal tuning switch open, observe and describe the mixer waveforms (t'=0) as a function of f, especially very close to $f=f_o$. Now turn on the integrator (t'=0.1 and 1 sec), repeat and interpret observations. Note that for f very close to f_o you will have to watch the output for a long time to fully describe the beating observed. Use a frequency counter to monitor the frequency f.

An effective bandwidth can be defined by measuring the slow beat amplitude for $f=f_o$, then going away from f_o and noting when the beat amplitude falls to $(1/2)^{1/2}$ of the initial amplitude. Do this for both values of t' and compare with part 1. In a narrow-band system such as this, you will need to get very close to f_o for the initial measurement and will therefore have to wait a while to define the beat amplitude. Can you interpret the observed bandwidth in terms of t'? Does it make any difference whether the signal tuning switch is open or closed? (A tuned preamp is used when noise signals at various frequencies away from f_o would overload the amplifier.) Does this lock-in respond to harmonics of f_o? Check this at $2f_o$ and $3f_o$. The result is most easily interpreted if you also look at the mixer waveform.

d. Phase-coherent lock-in action: Observe and interpret the output voltages for phase settings of 0 and 0, 90 degrees, and t' = 0, 0.1 and 1 sec.

The following are *extra* items to be done if time allows. Switch settings and input connections are to be decided by you.

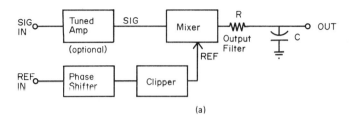

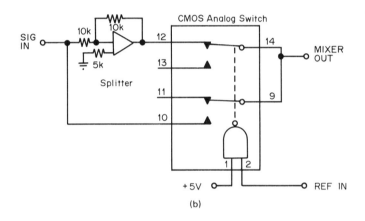

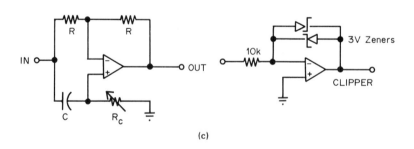

Figure 26.2 Lock-in using CMOS switch. (a) Overall block diagram; (b) Phase Detector or mixer; (c) Phase shifter; (d) Clipper.

LOCK-IN AMPLIFIER   26-5

e. When the reference is unbalanced, how does the mixer output change? What about the dc output for coherent or incoherent inputs?

f. The variable time-constant feature is a useful way of averaging random fluctuations in the coherent signal at f_o. Check this by using Osc 2 set very close to f_o so that the dc output beat is nearly stationary. Generate some noise (variation of signal amplitude) in the signal by banging on the oscillator or, better, summing in an incoherent signal.

The most useful feature of the lock-in is the ability to eliminate large nonrandom but incoherent signals not at f_o, recovering a signal buried in interference. Study this by feeding a signal which contains both f_o and some other frequency $f > 2f_o$ or $f < (f_o/2)$. This is done easily if one of the oscillators has a floating output so that the oscillator outputs can be added by connecting them in series. Otherwise, add the signals by feeding both Osc 2 and the Ref Output (phase = 0) to the input through 10 K Ohm resistors. Why does this provide the adder action? Observe the signal input to make sure the summing occurs. Observe both the mixer and dc output as the amplitude of Osc 2 is varied. You may have to turn off the LC tuning to take this measurement.

2. Lock-in Circuit II: CMOS Switching. We will use analog techniques instead. The switching is provided by a CMOS analog switch. Details are given in Fig. 26.2.

Examples of MOS switches are AD 7510, 7513, 7516 (Analog Devices); MC 14016 (Motorola); DG 200 (Siliconix); MM 451 and AH 0015 (National). Also included are an op amp phase shifter, whose output is constant in amplitude but continuously adjustable in phase. A tuned amplifier (see experiment 24) in the signal preamp is optional, but useful if the incoherent background is appreciable.

First perform the basic operations outlined for the diode lock-in above (nos. b and e are unnecessary since the IC is permanently balanced).

Because of its excellent characteristics, this circuit is capable of demonstrating a real-life signal recovery situation, such as the ones shown in text Fig. 18.12.

3. Coherent Detection of Light in the Presence of Large Ambient Noise (Optional project). A convincing example of a lock-in's ability to recover a coherent signal from larger incoherent background is shown in text Fig. 18.16. The photodetector amplifier may be simply constructed using a phototransistor, or a conventional op amp with a photodiode in the input. A light-emitting diode provides the coherent input. Incoherent background could be ambient light, or a flashlight. Such a circuit will detect the weak LED signal as long as the background level is not large enough to overload the photodetector or input amplifier.

The circuit uses a clever integrator subtraction technique to remove the ambient light (dc offset) from the photodiode signal applied to the input op amp. Any dc signal present in the output of A_1 acts as an input to integrator A_2, which ramps up or down until its output voltage cancels the dc input signal. (A_2 essentially establishes the ground point for the positive input of A_1.)

Test circuit operation initially with no ambient light by placing the LED photodiode circuit in a light-tight box. Remove the integrator feedback using the switch (SW). Using a strip of paper as a shutter, verify that as the coherent light intensity varies, the dc output also varies. If not, trace the coherent signal from its origin (D) to the LED driver (E) to the photodiode (A), amplifier (B), and mixer output (C with the capacitor removed).

Once this is all working, observe the output dc level using a digital multimeter as the box is opened. Does the dc level change when ambient light is admitted? Record the signal levels due to both coherent light and dc light at points A-E. (Limit the ambient light level so that no point in the circuit overloads.)

Now flip the switch so the integrator subtracts off the dc ambient light. Observe point A or B with a scope as you do so. Increase the intensity of the ambient light. Can a larger ambient intensity be tolerated as a result of the integrator offset subtraction? (A measure of the ambient signal is the voltage difference between point F and ground.)

EXPERIMENT 27: PHASE-LOCKED LOOPS

A. BACKGROUND

The phase-locked loop (**PLL**), though largely known in communications, also has instrumentation applications. In this experiment, you will explore basic PLL operation and then try a few applications: (a) FM demodulation, the basis for both linear and digital frequency shift data communications; (b) Divide-by-n, for creating sub-harmonics of a signal. For the brave, there is a project lab which gives you a reliable way of recording binary data on an ordinary tape recorder.

B. REFERENCES
Text, Chapter 18, Section 4.
See also, Signetics, Linear Applications Manual.

C. PROBLEMS
Text, Probs. 18.8, 18.9, and 18.10.

D. EQUIPMENT
Oscilloscope
Frequency counter
Voltage controlled oscillator (near the free running frequency of the PLL)
Digital multimeter
TTL binary counter (7490 or 7493) for divide-by-n.
A **split ± 5V supply** is recommended for the PLL.
565 phased-locked loop. See detailed specifications.

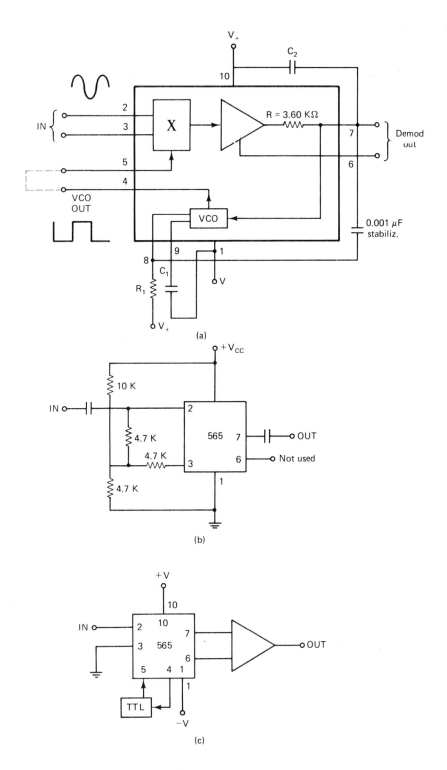

Figure 27.1 (a) Practical working circuit for 565 phase-locked loop. Block diagram with pin connections and external components; (b) Single-ended biasing; (c) Split power supply.

SIMPLIFIED DIAGRAM OF 565 VCO

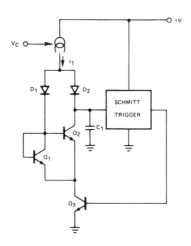

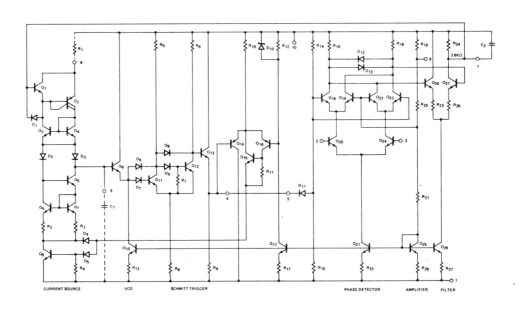

Figure 27.2 Internal circuit of the 565 phase-locked loop.

PHASE-LOCKED LOOPS   27-3

E. MEASUREMENTS

1. Basic 565 PLL Operations. Set up the circuit shown in Fig. 27.1(a). Try C_2 = .01 to 1 uF; V_+ = +5, V_- = -5; (or use single-ended 15 V). R_1 and C_1 set the free running frequency of the VCO; $f_o = 1/(3.7\ RC)$. Select f_o in the audio range. If a split power supply is not available, the modifications shown in Fig. 27.1(b) are required. The internal circuit is shown in Fig. 27.2.

Explore the free-running operation of the PLL, with no input signal. What is the free running frequency? How does it vary with R_1 or C_1? Select one combination and measure the frequency accurately. Leave R_1 and C_1 unchanged for the next section.

Measure the DC voltage out on pin 7 and the voltage difference between 6 and 7. Note that the value at pin 7 will vary depending on power supply, but the difference seen between 6 and 7 will not.

Insert a sine wave input at approximately the free running frequency and observe the VCO output. Is it tracking? (Trigger the scope using the input signal.) Measure the DC value of the demodulated output (pins 6 and 7). Now vary the input _frequency_, and observe the resulting change in DC output voltage, and in the _phase_ between input and VCO (see Eq. 15.15). What is the phase when the DC output is zero? What is the frequency then? (Compare: free running.)

Explore the lock and capture phenomenon. Note that both ranges depend on input signal _amplitude_; try several values from 50mV to 2V peak to peak. What can you say about the range of the DC output (pins 6-7) within the capture range?

2. FM Demodulation. Connect the VCO, driven by another signal in the audio range. That signal could be a sine wave or other function such as triangle or ramp. This provides the FM signal input. Note that the VCO portion of a second 565 PLL can be used.

Using this FM signal as an input, observe the demodulated output. Does it faithfully track the original signal? Note that C_2 will set an upper limit to the frequency at which the PLL can follow an input (Eq. 15.17). Note also that the <u>amplitude</u> of the modulation signal is also important. If too large, the FM input will make excursions outside of the lock range, and the PLL will be unable to follow a portion of the FM cycle.

This circuit is the basis for digital data transmission via phone lines. Adjust the two input frequencies to produce demodulated outputs of near 0 V (logic 0), and 1.0 V (logic 1). That signal can then drive TTL. If you arrange your input signal to drive a small speaker, and obtain the PLL input through a microphone, you can try digital data transmission via acoustic coupling on a telephone line.

3. Frequency Multiplication. Explore the possibility of frequency multiplication as shown in Fig. 18.21. The divide-by-N is inserted between pins 4 and 5 to the 565 PLL. Begin with simple powers of 2. Observe the waveforms at pins 4, 5, and 7. Note any frequency modulation at pin 4. Can you reduce it? Example waveforms are shown in Fig. 27.3.

4. Optional Project Lab: Digital Tape Recorder for Binary Data, see Fig. 27.4. Consult Signetics <u>Linear Applications Manual</u> for details.

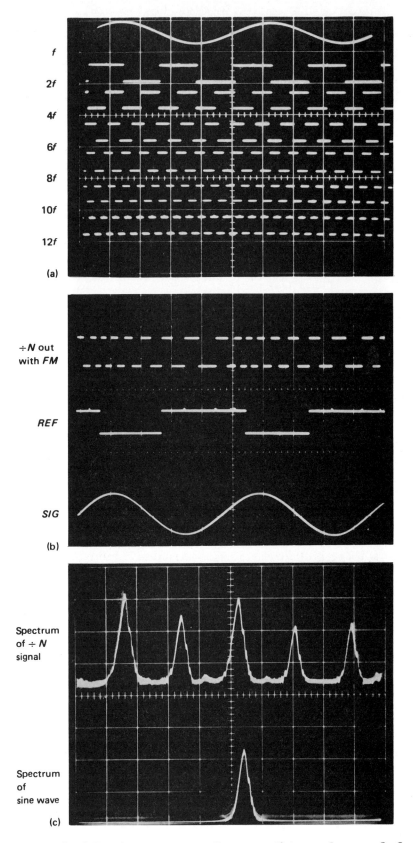

Figure 27.3 Divide-by-N waveforms (b) unplanned frequency modulation of the divided waveform (c) frequency spectrum of the divide-by-n output.

27-6   PHASE-LOCKED LOOPS

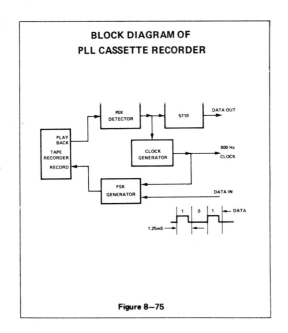
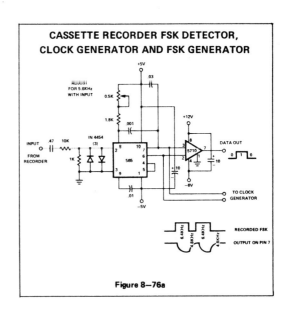
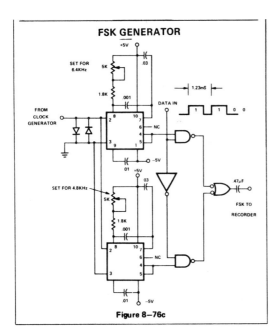

Figure 27.4 Digital data storage on an inexpensive cassette tape recorder, using PLL frequency-shift keying.

APPENDIX I: PIN DIAGRAMS FOR COMMON IC'S

Thanks to the following manufacturers for permission to reproduce the following pin diagrams from their data books:

Analog Devices Corporation: AD504 low noise op amp
    AD506 FET op amp
    AD520/521 instrumentation amplifier
    AD590 temperature sensor
    AD 533 multiplier
    AD7533 digital to analog converter
    AD7512 analog switch

Intel Corporation: 2114 1024x4-bit RAM, our suggested substitute when you get bored with the TTL 64-bit RAM.

Motorola Corporation: MC14433 dual slope A/D converter. See also the Intersil 7106 alternative.

RCA: CA3140 BiMOS op amp

Signetics Corporation, a subsidiary of U.S. Phillips Corporation: TTL pin diagram summary. The number designations are 54XX (extended temperature range) or 74XX (normal use), and LS stands for low-power Shottky TTL; the same pin numbers apply for ordinary TTL and for the CMOS equivalents specified by a 74XXC designation.
    NE5533/5534 low noise op amp
    LF355/356 BiFET op amps
    NE566 function generator
    NE/SE565 phase-locked loop

Texas Instruments: TL080 series BiFET op amps
    LM117, 217, 317 voltage regulators
    LM311 comparator
    SE556 dual precision timers

Note
LM and LF = National Semiconductor original part number
SE = Signetics original part number

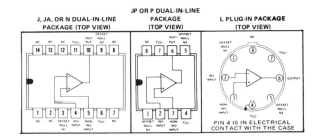

CA3140,

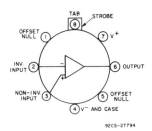

HIGH PERFORMANCE JFET INPUT OP AMPS

LF355/356

SINGLE AND DUAL LOW NOISE OP AMP

NE5533/5533A / NE/SE5534/5534A

COMMON FEATURES
(TYPICAL)
- Low input bias current 50pA
- Low input offset current 10pA
- High input impedance $10^{12}\Omega$
- Low input offset voltage 3mV
- Low V_{OS} temperature drift $5\mu V/°C$
- Low input noise current $0.01pA/\sqrt{Hz}$

APPLICATIONS
- Precision high speed integrators
- Fast A/D, D/A converters
- High impedance buffers
- Wideband, low noise, low drift amplifier

FEATURES
- Small-signal bandwidth: 10MHz
- Output drive capability: 600Ω, 10V (rms) at $V_S = \pm 18V$
- Input noise voltage: $4nV/\sqrt{Hz}$
- DC voltage gain: 100000
- AC voltage gain: 6000 at 10kHz
- Power bandwidth: 200kHz
- Slew-rate: $13V/\mu s$
- Large supply voltage range: ± 3 to $\pm 20V$

PIN CONFIGURATIONS

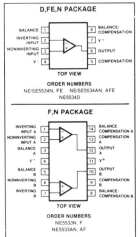

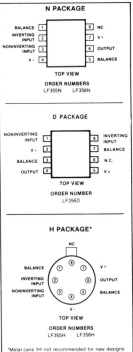

*Metal cans (H) not recommended for new designs

2114
1024 X 4 BIT STATIC RAM

	2114-2	2114-3	2114	2114L2	2114L3	2114L
Max. Access Time (ns)	200	300	450	200	300	450
Max. Power Dissipation (mw)	525	525	525	370	370	370

- **High Density 18 Pin Package**
- **Identical Cycle and Access Times**
- **Single +5V Supply**
- **No Clock or Timing Strobe Required**
- **Completely Static Memory**
- **Directly TTL Compatible: All Inputs and Outputs**
- **Common Data Input and Output Using Three-State Outputs**
- **Pin-Out Compatible with 3605 and 3625 Bipolar PROMs**

The Intel® 2114 is a 4096-bit static Random Access Memory organized as 1024 words by 4-bits using N-channel Silicon-Gate MOS technology. It uses fully DC stable (static) circuitry throughout — in both the array and the decoding — and therefore requires no clocks or refreshing to operate. Data access is particularly simple since address setup times are not required. The data is read out nondestructively and has the same polarity as the input data. Common input/output pins are provided.

The 2114 is designed for memory applications where high performance, low cost, large bit storage, and simple interfacing are important design objectives. The 2114 is placed in an 18-pin package for the highest possible density.

It is directly TTL compatible in all respects: inputs, outputs, and a single +5V supply. A separate Chip Select ($\overline{CS}$) lead allows easy selection of an individual package when outputs are or-tied.

The 2114 is fabricated with Intel's N-channel Silicon-Gate technology — a technology providing excellent protection against contamination permitting the use of low cost plastic packaging.

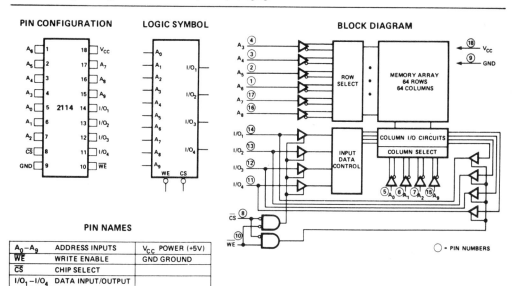

54LS/74LS83A 4-Bit Binary Full Adder, Fast Carry

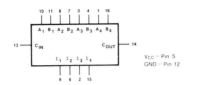

V_{CC} = Pin 5
GND = Pin 12

54LS/74LS181 4-Bit Arithmetic Logic Unit

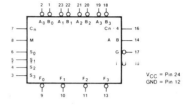

V_{CC} = Pin 24
GND = Pin 12

MODE SELECT - FUNCTION TABLE

MODE SELECT INPUTS				ACTIVE LOW INPUTS & OUTPUTS	
S_3	S_2	S_1	S_0	LOGIC (M = H)	ARITHMETIC[2] (M = L) (C_n = L)
L	L	L	L	$\bar{A}$	A minus 1
L	L	L	H	$\overline{AB}$	AB minus 1
L	L	H	L	$\bar{A}$ + B	$A\bar{B}$ minus 1
L	L	H	H	Logical 1	minus 1
L	H	L	L	$\overline{A + B}$	A plus (A + $\bar{B}$)
L	H	L	H	$\bar{B}$	AB plus (A + $\bar{B}$)
L	H	H	L	$\overline{A \oplus B}$	A minus B minus 1
L	H	H	H	A + $\bar{B}$	A + $\bar{B}$
H	L	L	L	$\bar{A}$B	A plus (A + B)
H	L	L	H	A $\oplus$ B	A plus B
H	L	H	L	B	AB plus (A + B)
H	L	H	H	A + B	A + B
H	H	L	L	Logical 0	A plus A*
H	H	L	H	$A\bar{B}$	AB plus A
H	H	H	L	AB	A$\bar{B}$ plus A
H	H	H	H	A	A

MODE SELECT INPUTS				ACTIVE HIGH INPUTS & OUTPUTS	
S_3	S_2	S_1	S_0	LOGIC (M = H)	ARITHMETIC[2] (M = L) (C_n = H)
L	L	L	L	$\bar{A}$	A
L	L	L	H	$\overline{A + B}$	A + B
L	L	H	L	$\bar{A}$B	A + $\bar{B}$
L	L	H	H	Logical 0	minus 1
L	H	L	L	$\overline{AB}$	A plus A$\bar{B}$
L	H	L	H	$\bar{B}$	(A + B) plus A$\bar{B}$
L	H	H	L	A $\oplus$ B	A minus B minus 1
L	H	H	H	A$\bar{B}$	A$\bar{B}$ minus 1
H	L	L	L	$\overline{A}$ + B	A plus AB
H	L	L	H	$\overline{A \oplus B}$	A plus B
H	L	H	L	B	(A + $\bar{B}$) plus AB
H	L	H	H	AB	AB minus 1
H	H	L	L	Logical 1	A plus A*
H	H	L	H	A + $\bar{B}$	(A + B) plus A
H	H	H	L	A + B	(A + $\bar{B}$) plus A
H	H	H	H	A	A minus 1

Definitions on back cover

54LS/74LS85 4-Bit Magnitude Comparator

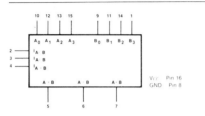

V_{CC} = Pin 16
GND = Pin 8

TRUTH TABLE 54LS/74LS85 4-Bit Magnitude Comparator

COMPARING INPUTS				CASCADING INPUTS			OUTPUTS		
A_3,B_3	A_2,B_2	A_1,B_1	A_0,B_0	$I_{A>B}$	$I_{A<B}$	$I_{A=B}$	A>B	A<B	A=B
$A_3>B_3$	X	X	X	X	X	X	H	L	L
$A_3<B_3$	X	X	X	X	X	X	L	H	L
$A_3=B_3$	$A_2>B_2$	X	X	X	X	X	H	L	L
$A_3=B_3$	$A_2<B_2$	X	X	X	X	X	L	H	L
$A_3=B_3$	$A_2=B_2$	$A_1>B_1$	X	X	X	X	H	L	L
$A_3=B_3$	$A_2=B_2$	$A_1<B_1$	X	X	X	X	L	H	L
$A_3=B_3$	$A_2=B_2$	$A_1=B_1$	$A_0>B_0$	X	X	X	H	L	L
$A_3=B_3$	$A_2=B_2$	$A_1=B_1$	$A_0<B_0$	X	X	X	L	H	L
$A_3=B_3$	$A_2=B_2$	$A_1=B_1$	$A_0=B_0$	H	L	L	H	L	L
$A_3=B_3$	$A_2=B_2$	$A_1=B_1$	$A_0=B_0$	L	H	L	L	H	L
$A_3=B_3$	$A_2=B_2$	$A_1=B_1$	$A_0=B_0$	L	L	H	L	L	H
$A_3=B_3$	$A_2=B_2$	$A_1=B_1$	$A_0=B_0$	X	X	H	L	L	H
$A_3=B_3$	$A_2=B_2$	$A_1=B_1$	$A_0=B_0$	H	H	L	L	L	L
$A_3=B_3$	$A_2=B_2$	$A_1=B_1$	$A_0=B_0$	L	L	L	H	H	L

Definitions on back cover

54LS/74LS89 16X4 Read/Write Memory, O/C

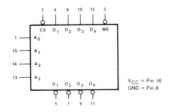

V_{CC} = Pin 16
GND = Pin 8

54LS/74LS89 16X4 Read/Write Memory, O/C

MODE SELECT—FUNCTION TABLES

OPERATING MODE	INPUTS			OUTPUTS
	$\overline{CS}$	$\overline{WE}$	D_n	$\bar{O}_n$
Write	L	L	L	H
	L	L	H	L
Read	L	H	X	Data
Inhibit Writing	H	L	L	H
	H	L	H	L
Store-Disable Outputs	H	H	X	H

Definitions on back cover

54LS/74LS95B 4-Bit Shift Register

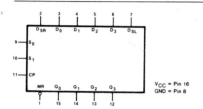

MODE SELECT—FUNCTION TABLE

OPERATING MODE	INPUTS							OUTPUTS			
	CP	MR	S_1	S_0	D_{SR}	D_{SL}	D_n	Q_0	Q_1	Q_2	Q_3
Reset (clear)	X	L	X	X	X	X	X	L	L	L	L
Hold (do nothing)	X	H	l	l	X	X	X	q_0	q_1	q_2	q_3
Shift Left	↑	H	h	l	X	l	X	q_1	q_2	q_3	L
	↑	H	h	l	X	h	X	q_1	q_2	q_3	H
Shift Right	↑	H	l	h	l	X	X	L	q_0	q_1	q_2
	↑	H	l	h	h	X	X	H	q_0	q_1	q_2
Parallel Load	↑	H	h	h	X	X	d_n	d_0	d_1	d_2	d_3

54LS/74LS194A 4-Bit Bidirectional Shift Register

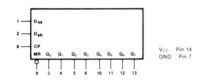

MODE SELECT—FUNCTION TABLES

OPERATING MODE	INPUTS					OUTPUTS			
	S	$\overline{CP}_1$	$\overline{CP}_2$	D_S	D_n	Q_0	Q_1	Q_2	Q_3
Parallel load	H	X	↓	X	l	L	L	L	L
	H	X	↓	X	h	H	H	H	H
Shift right	L	↓	X	l	X	L	q_0	q_1	q_2
	L	↓	X	h	X	H	q_0	q_1	q_2
Mode change	↑	L	X	X	X	no change			
	↑	H	X	X	X	undetermined			
	↓	X	L	X	X	no change			
	↓	X	H	X	X	undetermined			

Definitions on back cover

54LS/74LS164 8-Bit Serial-in-Parallel-out Shift Register

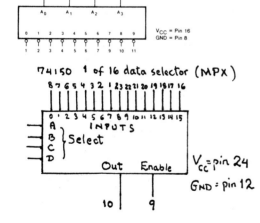

MODE SELECT—TRUTH TABLE

OPERATING MODE	INPUTS			OUTPUTS			
	$\overline{MR}$	D_{sa}	D_{sb}	Q_0	Q_1	–	Q_7
Reset (Clear)	L	X	X	L	L	–	L
Shift	H	l	l	L	q_0	–	q_6
	H	l	h	L	q_0	–	q_6
	H	h	l	L	q_0	–	q_6
	H	h	h	H	q_0	–	q_6

54LS/74LS42 BCD-To-Decimal Decoder

54LS/74LS151 8-to-1 Multiplexer

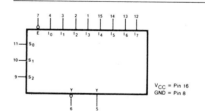

54LS/74LS154 4-to-16 Decoder/Demultiplexer

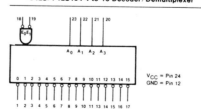

54LS/74LS90 Decade Ripple Counter

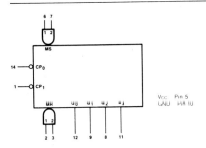

V_{CC} = Pin 5
GND = Pin 10

MODE SELECTION

RESET/SET INPUTS				OUTPUTS			
MR_1	MR_2	MS_1	MS_2	Q_0	Q_1	Q_2	Q_3
H	H	L	X	L	L	L	L
H	H	X	L	L	L	L	L
X	X	H	H	H	L	L	H
L	X	L	X	Count			
X	L	X	L	Count			
L	X	X	L	Count			
X	L	L	X	Count			

54LS/74LS93 4-Bit Binary Ripple Counter

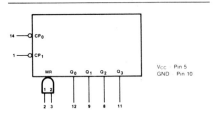

V_{CC} = Pin 5
GND = Pin 10

MODE SELECTION

RESET INPUTS		OUTPUTS			
MR_1	MR_2	Q_0	Q_1	Q_2	Q_3
H	H	L	L	L	L
L	H	Count			
H	L	Count			
L	L	Count			

54LS/74LS192 BCD Decade Up/Down Counter
54LS/74LS193 4-Bit Binary Up/Down Counter

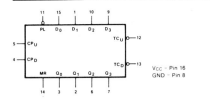

V_{CC} = Pin 16
GND = Pin 8

MODE SELECT — FUNCTION TABLE

OPERATING MODE	INPUTS					OUTPUTS		
	MR	$\overline{PL}$	CP_U	CP_D	D_0, D_1, D_2, D_3	Q_0, Q_1, Q_2, Q_3	$\overline{TC_U}$	$\overline{TC_D}$
Reset (clear)	H	X	X	L	X X X X	L L L L	H	L
	H	X	X	H	X X X X	L L L L	H	H
Parallel load	L	L	X	L	L L L L	L L L L	H	L
	L	L	X	H	L L L L	L L L L	H	H
	L	L	L	X	H X(b) X(b) H	$Q_n = D_n$	L	H
	L	L	H	X	H X(b) X(b) H	$Q_n = D_n$	H	H
Count up	L	H	↑	H	X X X X	Count up	H(b)	H
Count down	L	H	H	↑	X X X X	Count down	H	H(c)

NOTES
b. $\overline{TC_U}$ = CP_U at terminal count up (HXXH for "192" and (HHHH for "193")
c. $\overline{TC_D}$ = CP_D at terminal count down (LLLL)

Definitions on back cover

54LS/74LS190 BCD Decade Up/Down Counter
54LS/74LS191 4-Bit Binary Up/Down Counter

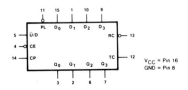

V_{CC} = Pin 16
GND = Pin 8

MODE SELECT — FUNCTION TABLE

OPERATING MODE	INPUTS				OUTPUTS	
	$\overline{PL}$	$\overline{U}/D$	$\overline{CE}$	CP	D_n	Q_n
Parallel load	L	X	X	X	L	L
	L	X	X	X	H	H
Count up	H	L	l	↑	X	count up
Count down	H	H	l	↑	X	count down
Hold "do nothing"	H	X	H	X	X	no change

54LS/74LS160 Synchronous 4-Bit Decade Counter
54LS/74LS161 Synchronous 4-Bit Binary Counter

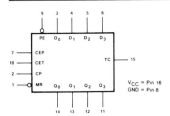

V_{CC} = Pin 16
GND = Pin 8

MODE SELECT — FUNCTION TABLE

OPERATING MODE	INPUTS						OUTPUTS	
	$\overline{MR}$	CP	CEP	CET	$\overline{PE}$	D_n	Q_n	TC
Reset (Clear)	L	X	X	X	X	X	L	L
Parallel Load	H	↑	X	X	l	l	L	L
	H	↑	X	X	l	h	H	(b)
Count	H	↑	h	h	h	X	count	(b)
Hold (do nothing)	H	X	l	X	h	X	q_n	(b)
	H	X	X	l	h	X	q_n	L

54LS/74LS221 Dual Monostable Multivibrator

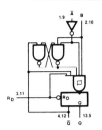

One-Shots & R-S Latch

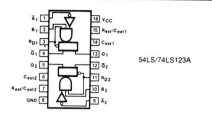

54LS/74LS123A

FUNCTION TABLE

INPUTS			OUTPUTS	
$\overline{R}_D$	$\overline{A}$	B	Q	$\overline{Q}$
L	X	X	L	H
X	H	X	L	H
X	X	L	L	H
H	L	↑	⊓	⊔
H	↓	H	⊓	⊔
↑	L	H	⊓	⊔

H = HIGH voltage level
L = LOW voltage level
X = Don't care
↑ = LOW-to-HIGH transition
↓ = HIGH-to-LOW transition

54LS/74LS75 Quad Bistable Latch

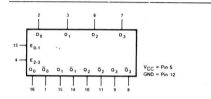

V_{CC} = Pin 5
GND = Pin 12

54LS/74LS273 Octal "D" Flip-Flop

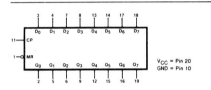

V_{CC} = Pin 20
GND = Pin 10

Flip-Flops

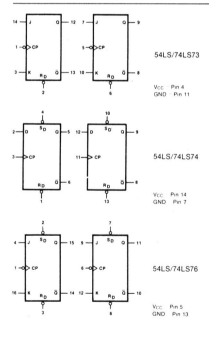

54LS/74LS73
V_{CC} = Pin 4
GND = Pin 11

54LS/74LS74
V_{CC} = Pin 14
GND = Pin 7

54LS/74LS76
V_{CC} = Pin 5
GND = Pin 13

Flip-Flops

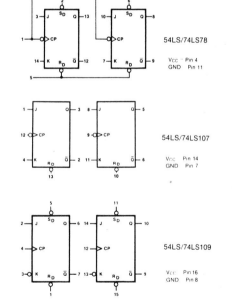

54LS/74LS78
V_{CC} = Pin 4
GND = Pin 11

54LS/74LS107
V_{CC} = Pin 14
GND = Pin 7

54LS/74LS109
V_{CC} = Pin 16
GND = Pin 8

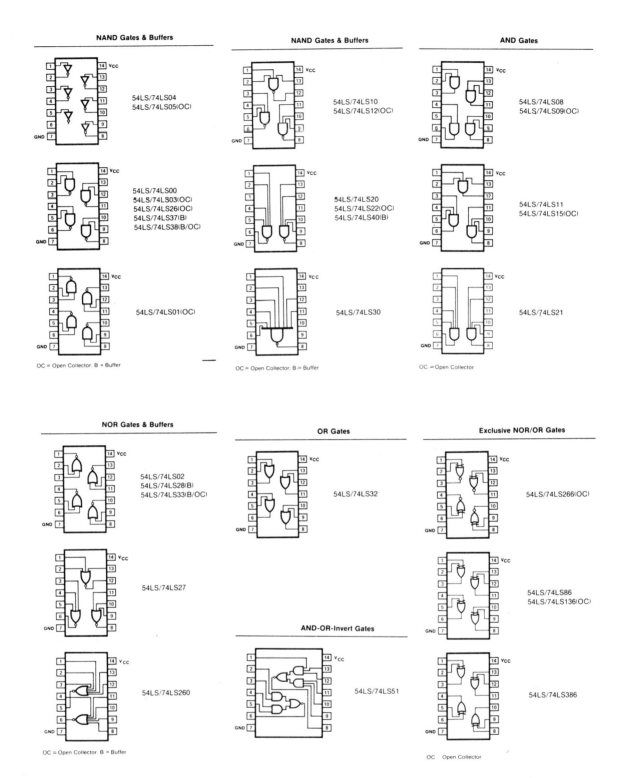

Am2502/2503/2504

LOGIC DIAGRAM/SYMBOLS

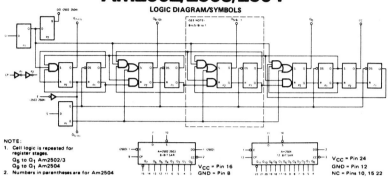

NOTE:
1. Cell logic is repeated for register stages.
   Q_5 to Q_1 Am2502/3
   Q_9 to Q_1 Am2504
2. Numbers in parentheses are for Am2504.

Am2502/2503 8-BIT SAR V_{CC} = Pin 16 GND = Pin 8

Am2504 12-BIT SAR V_{CC} = Pin 24 GND = Pin 12 NC = Pins 10, 15, 22

MC14433

CMOS LSI
(LOW-POWER COMPLEMENTARY MOS)

3½ DIGIT A/D CONVERTER

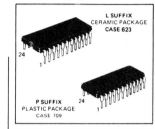

L SUFFIX
CERAMIC PACKAGE
CASE 623

P SUFFIX
PLASTIC PACKAGE
CASE 709

ORDERING INFORMATION

MC14XXX — Suffix Denotes
- L Ceramic Package
- P Plastic Package

- Accuracy: ±0.05% of Reading ±1 Count
- Two Voltage Ranges: 1.999 V and 199.9 mV
- Up to 25 Conversions/s
- Z_{in} > 1000 M ohm
- Auto-Polarity and Auto-Zero
- Single Positive Voltage Reference
- Standard B-Series CMOS Outputs—Drives One Low Power Schottky Load
- Uses On-Chip System Clock, or External Clock
- Low Power Consumption: 8.0 mW typical @ ±5.0 V
- Wide Supply Range: e.g., ±4.5 V to ±8.0 V
- Overrange and Underrange Signals Available
- Operates in Auto Ranging Circuits
- Operates with LED and LCD Displays
- Low External Component Count

BLOCK DIAGRAM

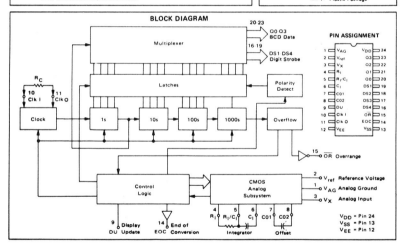

V_{DD} = Pin 24
V_{SS} = Pin 13
V_{EE} = Pin 12

TYPES TL080, TL081, TL085, GENERAL PURPOSE BIFET SERIES

38 devices cover commercial, industrial, and military temperature ranges
- Low Power Consumption
- Wide Common-Mode and Differential Voltage Ranges
- Low Input Bias and Offset Currents
- Output Short-Circuit Protection
- High Input Impedance...JFET-Input Stage
- Internal Frequency Compensation (Except TL080, TL080A)
- Latch-Up-Free Operation
- High Slew Rate...13 V/μs Typ
- Low Input Offset Voltage....0.5 mV on TL087 and 2 mV on TL088

The TL081 family of BIFET op amps feature eight devices — 4 singles, 2 duals, and 2 quads. The circuit simplicity, small chip size, and ease of manufacture result in high production yields and low cost to the user.

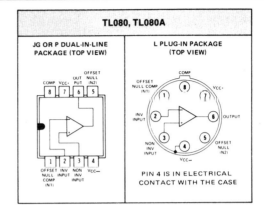

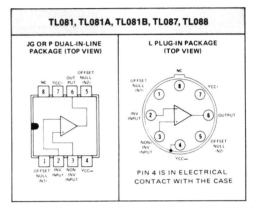

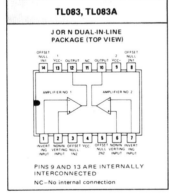

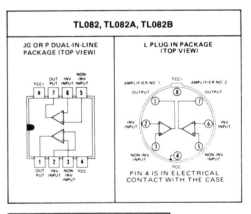

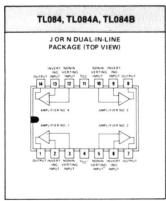

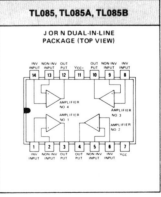

PIN CONFIGURATIONS
Top View

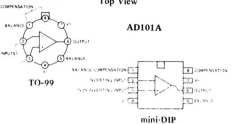

AD101A

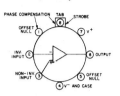

CA3130,

LOW DRIFT, LOW NOISE OP AMP
AD504

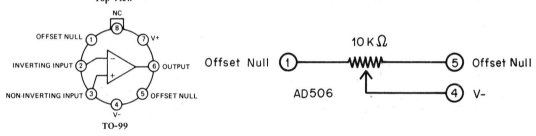

HIGH ACCURACY FET-INPUT OP AMP
AD506

PIN CONFIGURATION

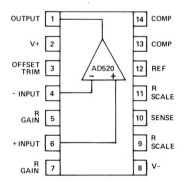

PIN CONFIGURATION

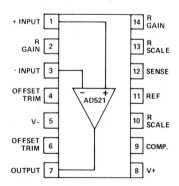

TYPES LM117, LM217, LM317
3-TERMINAL ADJUSTABLE REGULATORS

- Output Voltage Range Adjustable
- Guaranteed I_O Capability of 1.5 A for TO-220 package, 500 mA for LA and TO-202 packages
- Input Regulation Typically 0.01%/V Input Change.
- Output Regulation Typically 0.1%
- Peak Output Current Constant Over Temperature Range of Regulator
- Popular 3-Lead Packages
- Ripple Rejection Typically 80 dB

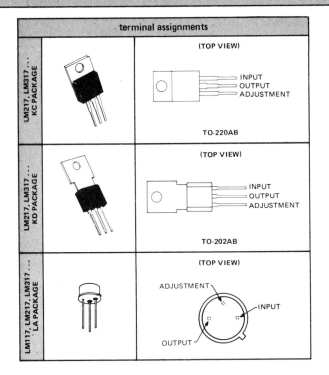

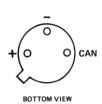

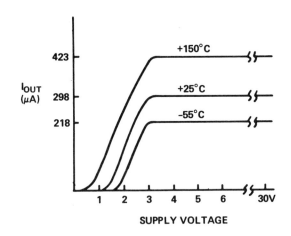

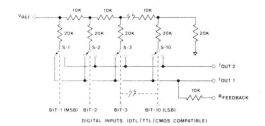

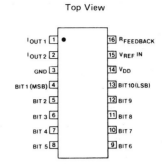

Top View

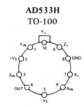

AD533H
TO-100

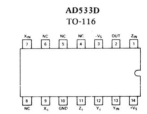

AD533D
TO-116

TOP VIEW

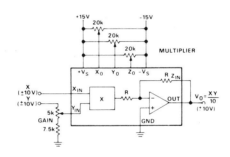

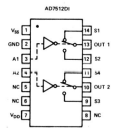

TYPES LM111, LM311
SINGLE DIFFERENTIAL COMPARATORS WITH STROBE

- Fast Response Times...115 ns Typ
- Strobe Capability
- Designed to be Interchangeable with National Semiconductor LM111 and LM311
- Maximum Input Bias Current...300 nA
- Maximum Input Offset Current...70 nA
- Can Operate From Single 5-V Supply

TIMER

NE/SE555/SE555C

SE555F,H,N,N-14 • SE555C,F,H,N,N-14 • NE555F,H,N,N-14

FEATURES
- Turn off time less than 2µs
- Maximum operating frequency greater than 500kHz
- Timing from microseconds to hours
- Operates in both astable and monostable modes
- High output current
- Adjustable duty cycle
- TTL compatible
- Temperature stability of 0.005% per °C
- SE555 Mil std 883A,B,C available M38510 (JAN) approved, M38510 processing available.

PIN CONFIGURATIONS

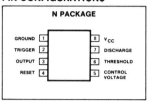

APPLICATIONS
- Precision timing
- Pulse generation
- Sequential timing
- Time delay generation
- Pulse width modulation
- Pulse position modulation
- Missing pulse detector

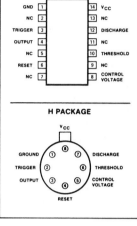

PIN CONFIGURATION

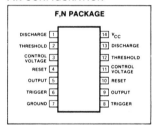

BLOCK DIAGRAM

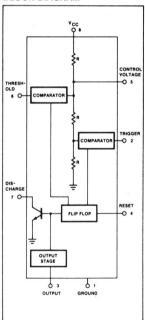

BLOCK DIAGRAM

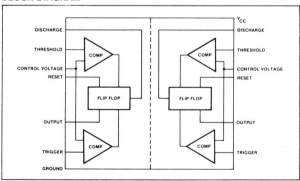

FUNCTION GENERATOR

NE/SE566

NE/SE566-F,N,T

DESCRIPTION
The SE/NE 566 Function Generator is a voltage controlled oscillator of exceptional linearity with buffered square wave and triangle wave outputs. The frequency of oscillation is determined by an external resistor and capacitor and the voltage applied to the control terminal. The oscillator can be programmed over a ten to one frequency range by proper selection of an external resistance and modulated over a ten to one range by the control voltage, with exceptional linearity.

FEATURES
- Wide range of operating voltage (up to 24 volts)
- High linearity of modulation
- Highly stable center frequency (200 ppm/°C typical)
- Highly linear triangle wave output
- Frequency programming by means of a resistor or capacitor, voltage or current
- Frequency adjustable over 10 to 1 range with same capacitor

APPLICATIONS
- Tone generators
- Frequency shift keying
- FM modulators
- Clock generators
- Signal generators
- Function generators

PIN CONFIGURATIONS

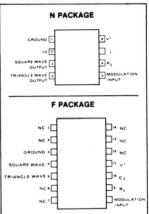

BLOCK DIAGRAM

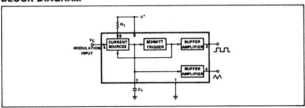

PHASE LOCKED LOOP

NE/SE565

NE/SE565-F,K,N

DESCRIPTION
The SE/NE565 Phase-Locked Loop (PLL) is a self-contained, adaptable filter and demodulator for the frequency range from 0.001Hz to 500kHz. The circuit comprises a voltage-controlled oscillator of exceptional stability and linearity, a phase comparator, an amplifier and a low-pass filter as shown in the block diagram. The center frequency of the PLL is determined by the free-running frequency of the VCO; this frequency can be adjusted externally with a resistor or a capacitor. The low-pass filter, which determines the capture characteristics of the loop, is formed by an internal resistor and an external capacitor.

PIN CONFIGURATIONS

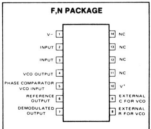

FEATURES
- Highly stable center frequency (200ppm/°C typ.)
- Wide operating voltage range (±6 to ±12 volts)
- Highly linear demodulated output (0.2% typ.)
- Center frequency programming by means of a resistor or capacitor, voltage or current
- TTL and DTL compatible square-wave output; loop can be opened to insert digital frequency divider
- Highly linear triangle wave output
- Reference output for connection of comparator in frequency discriminator
- Bandwidth adjustable from < ±1% to > ±60%
- Frequency adjustable over 10 to 1 range with same capacitor

BLOCK DIAGRAM

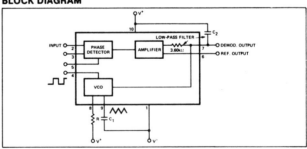

APPENDIX II. How To Identify Resistors and Capacitors

Resistor color code is a series of colored bands

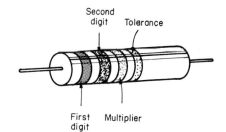

Resistor Color Code

Color	Number
Black	0
Brown	1
Red	2
Orange	3
Yellow	4
Green	5
Blue	6
Violet	7
Gray	8
White	9

The code shown to the left is for ordinary (low precision) resistors. Occasionally, capacitors are also found marked with the same code.

Capacitor Code Markings

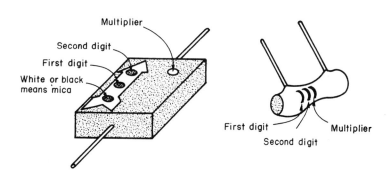

APPENDIX III: SOURCES OF SUPPLY

Sources of supply are listed in several categories:

1. Companies (e.g., Jameco) which cater to hobbyists, schools, small buyers, often selling at a considerable discount price. Beware however of "cosmetic rejects" or parts with manufacturer unlabeled - below specification operation may be observed. Hobbyist suppliers generally carry a limited subset of IC's, generally including only the most popular types.

2. Distributors (e.g., Hamilton Avnet or Almac Stroum) which cater to professional customers and often require a sizeable minimum order. Distributors have the widest selection of both manufacturers and of part numbers, and are also the place to get invaluable manufacturers' handbooks at minimum cost.

3. Manufacturers themselves usually don't want to deal with the small purchaser, especially not directly to the factory. On the other hand, by writing to the address listed you can get the addresses of both distributors and a category called the manufacturer's regional representative. Some manufacturers (e.g., Intel and Intersil) have a liberal policy towards educational institutions, with free literature and low priced parts kits or evaluation kits.

HOBBYIST OR DISCOUNT SUPPLY HOUSES

Alba Electronics (formerly Continental Specialties Corp.)
23 Alba Street   800-243-6953 (toll free)
New Haven, Conn. 06058   203-467-5590

Advanced Computer Products   800-854-8230 (toll free)
P.O. Box 17329
Irvine, CA 92713

Anacrona   213-641-4064
P.O. Box 2208
Culver City, CA 90230

Digi-Key Corporation   218-681-6674
P.O. Box 677
Thief River Falls, MN 56701

E&L Instruments   203-735-8774
61 First St.
Derby, Conn. 06418

Electronic Systems   408-448-0800
P.O. Box 21638
San Jose, CA 95151

Hobbyworld Electronics, Inc.   800-423-5387 (toll free)
19511 Business Center Drive   213-886-9200
Northridge, CA 91324

Jade Computer Products   213-679-3313
4901 West Rosecrans
Hawthorne, CA 90250

Mini Micro Mart, Inc.   315-422-4467
1618 James Street
Syracuse, NY 13203

Priority One Electronics   800-423-5922 (toll free)
9161 Deering Ave.   213-709-5464
Chatsworth, CA 91311

MANUFACTURERS

Advanced Micro Devices
901 Thompson Place
Sunnyvale   CA   94086   (408) 732-2400

American Microsystems, Inc.
3800 Homestead Road
Santa Clara   CA   95051   (408) 246-0330

Analog Devices
Route 1 Industrial Park
Norwood   MA   02062   (617) 329-4700

Analogic Corporation
Audobon Road
Wakefield   MA   01880   (617) 246-0300

Beckman Instruments
2500 Harbor Boulevard
Fullerton   CA   92634   (714) 871-4848

Burr-Brown
International Airport Park
Tucson   AZ   85734   (602) 746-1111

Cherry Semiconductor Corp.
2000 South County Trail
East Greenwich   RI   02818   (401) 885-3600

EXAR Integrated Systems
750 Palomar Avenue
Sunnyvale   CA   94088   (408) 732-7970

Fairchild
464 Ellis Street
Mountain View   CA   94043   (415) 962-5011

Intel
3065 Bowers Avenue
Santa Clara   CA   95051   (408) 987-8080

Intersil
10710 No. Tantau Avenue
Cupertino   CA   95014   (408) 996-5000

Jameco Electronics
1355 Shoreway Road
Belmont   CA   94002   (415) 592-8097

Lambda Semiconductor
121 International Drive
Corpus Christi   TX   78410   (512) 883-6251

MOS Technology
Valley Forge Corp. Center
Norristown   PA   19403   (215) 666-7950

Mostek
1215 W. Crosbby Road
Carrollton   TX   75006   (214) 323-6000

Motorola
5005 E. McDowell Road
Phoenix   AZ   85008   (602) 244-6900

National Semiconductor
2900 Semiconductor Drive
Santa Clara   CA   95051   (408) 737-5000

Precision Monolithics
1500 Space Park Drive
Santa Clara   CA   95050   (408) 245-9222

RCA
Box 3200
Somerville   NJ   08876   (201) 685-6000

Signetics
811 E. Arques Avenue
Sunnyvale CA 94086   (408) 739-7700

Siliconix
2201 Laurelwood Road
Santa Clara CA 95054   (408) 988-8000

Texas Instruments
PO Box 225012
Dallas TX 75265   (214) 238-6611

Teledyne Philbrick
Allied Drive
Dedham MA 02026   (617) 329-1600

Zilog, Inc.
1315 Dell Avenue
Campbell CA 95008   (408) 446-4666

NOTES

NOTES

NOTES

NOTES

NOTES

NOTES

NOTES

NOTES

NOTES

NOTES

NOTES

NOTES